A Naturalist Buys an Old Farm

Books by Edwin Way Teale

GRASSROOT JUNGLES
THE GOLDEN THRONG
NEAR HORIZONS
DUNE BOY
THE LOST WOODS
DAYS WITHOUT TIME
NORTH WITH THE SPRING
JOURNEY INTO SUMMER
AUTUMN ACROSS AMERICA
WANDERING THROUGH WINTER
CIRCLE OF THE SEASONS
A WALK THROUGH THE YEAR
SPRINGTIME IN BRITAIN
ADVENTURES IN NATURE
THE JUNIOR BOOK OF INSECTS
THE STRANGE LIVES OF FAMILIAR INSECTS
PHOTOGRAPHS OF AMERICAN NATURE

Books Edited by Edwin Way Teale

WALDEN
GREEN TREASURY
AUDUBON'S WILDLIFE
THE THOUGHTS OF THOREAU
THE INSECT WORLD OF J. HENRI FABRE
THE WILDERNESS WORLD OF JOHN MUIR

EDWIN WAY TEALE

A Naturalist Buys an Old Farm

with a new introduction by
Ann Haymond Zwinger

Bibliopola Press
UConn Co-op, Storrs, CT.
Distributed by the University Press of New England

To
the three who helped
us find Trail Wood

JAMES A. SLATER
WENDELL DAVIS
ALISON DAVIS

Teale Archives, University of Connecticut
Originally published by Dodd, Mead & Company, Inc.
Reissued 1998, with a new forward by Ann Haymond Zwinger
by Bibliopola Press, UConn Co-op, Storrs, CT, 06269-2019
Distributed by the University Press of New England,
Hanover, N.H. 03755-2048
Manufactured in the United States

Library of Congress Catalog Card Number
98-73453

ISBN 0-939883-02-3

Cover design by Wendell Minor
Interior woodcuts by Michael McCurdy

FOREWORD

I remember vividly the first time I drove up the lane at Trail Wood. The impression is so lasting that I need only close my eyes to bring it back in all its charm and essence. My feelings upon entering that two-track way must have been the same as those of dozens of other visitors, before and after me. The change came the moment I turned off the black, tire-humming asphalt road onto the unmarked way into an immediate, other-worldly serenity. Trees framed the passageway. Wildflowers brightened the thick undergrowth of green. Water twinkled in Hampton Brook. No habitations were visible. Buffers of hardwoods, screens of willows, verdant berms shielded Trail Wood.

And, at the end of the narrow lane, behind the big bed of Nellie's glorious day lilies, the story-book white house perched on a rise. Beyond was the charming pond overlooked by the writing cabin Edwin built, a faithful copy of Henry David Thoreau's at Walden Pond.

That first memory washed over me every time I returned over the years, pleasure in the familiar, curious about what was in bloom this time, what quick creature would flee across the road or flit from branch to branch above my head, the eternal enchantment of a Connecticut Camelot, an almost mystical sense of entering an enchanted place.

And so must Edwin and Nellie have felt when they drove up the lane on June 11, 1959, the new owners of this paragon of a place. *A Naturalist Buys an Old Farm* tells how they found Trail Wood, and it's a story worth telling for it delineates Edwin's character and determination and the yearning both had to be closer to a natural world that held their interest and devotion.

For thirty years Edwin and Nellie lived in Baldwin, Long Island, where initially Edwin commuted to the city to work as a magazine writer for "Popular Science." In his spare time he wrote about insects and photographed them, a hobby that he pursued with imagination and skill, immobilizing insects in the refrigerator, removing them with a background all in place, snapping the shutter just as the insect revived. His work was innovative and soon attracted the attention of publishers.

When he felt that he had established a secure income for his family, he cut his ties with the magazine and cast his lot with freelance writing and photography. His "Independence Day" came on October 16, 1941, and in the years that I knew him, he never failed to make note of it. Many splendid books followed through the years. And in that time came a growing realization that

Baldwin had become crowded. That this was not where they wished to live. And that understanding must have become unbearably poignant with the loss of their only son, David, in World War II; surely the house in Baldwin was filled with heartbreaking memories. They found themselves living in a world too much with them.

That the years preceding Trail Wood had entailed so much travel may also have had something to do with their wish for a quiet space in which to settle, to explore, to work. Beginning in 1947, Edwin undertook to write four volumes about the seasons across America, a huge project. The Teales traveled across and back, and up and down the United States. The trips required impeccable planning, arduous hours of research and interviewing, to say nothing of endless hours of driving. Yet every night Edwin unlimbered his portable typewriter and typed his day's handwritten field notes. In the introduction to each book he tallies the impressive number of days he and Nellie spent on the road.

Doubtless Edwin and Nellie had thought about leaving Baldwin, but not until he'd completed the volumes on spring, autumn, and summer, did they actively and methodically begin to search more seriously for a more compatible place to live. How Edwin and Nellie found the original seventy-nine acres is totally typical of Edwin's methodical temperament. He kept daily journals that listed at the beginning of the month what he wished to accomplish, with a corresponding list at the end of what he did accomplish. His lists of writing tasks, posted on the bulletin board in his office, were crossed through in bright colors as each task was finished.

Edwin applied this tidy way to finding Trail Wood: as he tells in *A Naturalist Buys an Old Farm*, he drew widening circles on a map from a focus point of New York where many of their friends and much of Edwin's professional help lived, and systematically searched these rings for a place where they could settle and live as they wished to live. They'd find the house they liked but the land was lacking, or vice versa, but with the help of friends, they discovered a place that had both land and living quarters and fulfilled all their requirements. Both at the beginning of their sixth decade, they had finally found home.

That they found these acres was pure serendipity, but then that too was typical of Edwin. Once he began a project, he completed it usually 200 percent to his satisfaction. When they found Trail Wood, they must have sensed that there were enough acres here to protect them and, perhaps more important, that there were enough acres to explore so that they would never run out of walks to take, discoveries to make, birds to sight, animals to observe. With their incredible curiosity and intelligence they needed to find always one more flower or a rare moss (who's ever heard of Schreber's moss?) or new fern — Edwin was proud of the fact that the variety of habitats at Trail Wood gave them half the recorded number of ferns for Connecticut.

They celebrated their 50th wedding anniversary in the white story-book house. Nellie, a fabulous reader, read out loud in front of the big fieldstone fireplace nearly every night of their lives, books they enjoyed or Edwin's manuscripts (a tremendous help for him). Excellent birders, binoculars always lay on the window sill, handy. Here they read *Don Quixote*, the book they both loved and that they read whenever times were difficult, when there were more problems than solutions, when they were discouraged or dismayed.

These must have been the most rewarding years of their lives. It was here that he finished the fourth of the seasons books, *Wandering through Winter.* And it was up the lane at Trail Wood that Edwin wheeled a wheelbarrow full of "New York Times Book Reviews" that praised its selection as winner of a Pulitzer Prize, the first book of natural history ever so honored.

Edwin began *A Naturalist Buys an Old Farm* with a chapter title as it might have been written by Cervantes. To begin a book in this way somehow forecasts that any tribulations were over, that the old farm was a new beginning. And indeed it was. If there was ever a book full of Edwin's particular and delightful nature writing, this book of discovery is it. Almost every page is a walk he took, places he and Nellie explored, hours spent in a hammock observing cheeky squirrels and rotund woodchucks and counting chipmunk chirps, or the relish of spring's first bloodroot along the lane.

Edwin was already an established writer when he wrote *A Naturalist Buys an Old Farm* but he might as well have been a beginner for all the wonderful treasures that he found. He and Nellie immediately set about exploring, laying out wildflower walks, watching fireflies, naming — Firefly Meadow, North Woods, Wild Apple Glades — and marking paths, following the seasons through brush and meadow and hence the name they chose for where they lived: Trail Wood.

Many other writers have written about the areas in which they live, and many of the books are memorable as were the journals of Thoreau, Edwin's muse. But most of these books concentrate on larger areas, often on land not personally owned (I use "own" simply to emphasize a place to which you have constant access, where you commonly sleep and carry out your days, that you can explore at leisure). Edwin's relationship with Trail Wood was close focus. He knew it intimately because he could observe the dawns and sunsets. This wonderful potpourri of a book would have been impossible could he not have walked the land's bounds on a regular basis and reveled in a totally sequestered, self-contained landscape because he *lived* there.

A Naturalist Buys an Old Farm is more than a love letter to a place; it's a recognition of a life of accomplishment on many levels: professional, personal, emotional. The enchantment of Trail Wood is such that it took me a long time to understand that although the land was natural, it was not pristine. Edwin was aware that their tenancy was only temporary in the time span of the land's existence.

Edwin was interested in local history and knew that Trail Wood had been grazed and farmed. He understood that changes in the land are inevitable, part of them natural changes through time, part of them human changes through plan. Edwin felt a sense of continuation, of belonging here for a period of the land's history, and in that communion he found comfort and great fascination, recognizing that the land was here first and would continue to be after all its occupants were gone, a persistent, eternal presence, able to change but also able to endure.

The presence of Edwin and Nellie is appreciable. Edwin had the pond excavated the year after they bought Trail Wood. He piled the stones for walls along Hampton Brook and fought a running battle with beavers there. He added his own "writing cabin" by the pond, and the little screened gazebo where they often had supper. He judiciously cut trees to maintain the meadows and enrich the variety of habitats for plants and animals.

But they always saw themselves as part of the land, and the changes they made were like their personalities, compatible and sympathetic. Their human stewardship, their care and solicitude remains as much a part of Trail Wood as the kind of rocks, the dragonflies, the warblers, and the woodcocks for which they eagerly listened each spring. Still, the Trail Wood visitors know today is different from Edwin and Nellie's Trail Wood, which was different from what the land was when they moved there. Change is the one constant.

I think the unspoken unifier of my relationship with Nellie and Edwin, more than working on a book together, was that each of us had found our own "home place," for them in Connecticut, for us in Colorado. From my place in the Colorado Rockies came my first book and my early understandings of place and time and aspen trees. *A Naturalist Buys an Old Farm* was far from Edwin's first book, but it had the same enthusiasm of discovery and pleasure as if it were.

In the widest sense, *A Naturalist Buys an Old Farm* depicts a human need to belong to a piece of earth and, for the naturalist, to learn it, to try to understand it. The bountifulness of Trail Wood enveloped Nellie and Edwin in peace. Trail Wood was more than a plot of ground, it was a world that they wrapped about their shoulders to keep out the chill. When Edwin wrote about their search and their finding, he wrote for everyone who has that yearning for permanent home. With *A Naturalist Buys an Old Farm* he wrote a guidebook to life, liberty, and the importance of happiness. *A Naturalist Buys an Old Farm* is classical natural history but it is also more: it is a classical understanding of man's need to have a stable base, a place to protect and to celebrate, a place to go out from but always to come home to. Knowing how and why this land came into their possession helps us to understand the tremendous meaning that it held for Edwin and Nellie, the incredible beauty not only of this land, but the whole earth upon which they lived.

All of us who have a place dear to our hearts, that anchors our lives, recognize Edwin's story and understand more about ourselves. His orderly finding and settling at Trail Wood gives us a kind of faith in orderly time and procedure that we need these chaotic days. *A Naturalist Buys an Old Farm* is a guidebook to contentment, a how-to book on living well and lightly with the land.

Trail Wood had its own magic, yes, but Edwin's books, *A Naturalist Buys an Old Farm* and, later, *A Walk through the Year*, intensified that magic by helping us outsiders to see what he saw, recorded in his graceful prose. His details of Trail Wood's delights lured visitors, by ones and twos and threes, adding up to thousands over the years. Through Edwin's and Nellie's generosity, Trail Wood became more available to the public through their donation of the land and house to the Connecticut Audubon Society.

It may be that one needs never actually see Trail Wood: the portraits that Edwin gives a reader are more than enough and, what's more, we see it through his keen and observant eye as we could never see it on our own. That this Shangri-la exists may be all we need to know.

Edwin died in 1980, two days after the thirty-ninth anniversary of his own "Independence Day." Nellie lived thirteen more years, most of it at Trail Wood, and chose to die on the anniversary of Edwin's death. I once wrote that I could not think of them separately although they were distinctly different personalities; it was the melding that made me always think of them as "NellieandEdwin."

Their hours at Trail Wood were rewarding, productive, and happy ones, and this is the legacy that they have left me, as well as for today's visitor, a benediction of their dear understanding, and of time that runs not by calendars but by the sun and the moon and by the celebration of seasons and solstices.

Ann Haymond Zwinger

Contents

CHAPTER ONE

Three Circles on a Map

If this were a chapter in Cervantes' *Don Quixote*, it might well be entitled: "Wherein is Discovered the Surroundings in Which we Find Ourselves and Other Passages Worthy of Happy Memory." It recounts our search for a home in the country. It tells where we are and how we got there.

A friend of mine once asked me if it were true—as he had heard—that we placed a silver dollar on a map of this state and moved it about until we found a place where no communities of any size were under the coin and then settled there. That story, no doubt, had its source in the three red circles I traced on a map of the Eastern states.

For many years in my life making a living pulled me toward the city and living as I wanted to live pulled me toward the country. When Nellie and I moved east from the Midwest, we settled in a pleasant village on the edge of the country, twenty-two miles from Manhattan on the south shore of Long Island. Baldwin, in those days, held extensive patches of woodland and wild sea meadows stretched out to the bay. During the succeeding three decades, while I was working first in a magazine office in the city and then as a freelance author beginning my nature books, we saw our surroundings change. The open countryside, the stretches of woodland, the sea meadows all contracted or disappeared. They were buried beneath subdivisions and developments, buildings and more buildings. Noise and congestion increased. We saw all the small towns around lose their individuality. They merged into one vast extension of Brooklyn. One by one the roots that held us were cut. When personal responsibilities no longer anchored us there, we began looking for our longed-for home in the country.

At first, realizing we could move wherever we wished, we thought of distant places we had enjoyed in our travels—the Oregon coast, the Colorado Rockies, Minnesota's land of lakes, California's redwood country, and Monterey Peninsula. However such a move would mean leaving our friends behind. It would mean losing contact with familiar centers of library and museum research in New York. In the end we decided to find a country home somewhere within a day's reach of the city. With New York as a center, I traced with a red pencil three concentric circles on a map of the Eastern states. One circle had a radius of fifty miles, the second 100 miles, the third 150 miles. Then, in the territory in the circular corridors enclosed within these lines, we began our three-year search.

We made a list of all the things we hoped to have on our place in the country: a wood, a stream, a swamp, open meadows, a Cape Cod-type farmhouse, a fireplace, at least fifty acres of ground. It seemed an impossible list. But whenever we had free time, we continued our search. We explored up the Hudson River Valley; we wandered down side roads in western Connecticut. We visited the Pine Barrens and the New Jersey coast. We went out with real estate men. We kept track of advertisements—learning to decode their flamboyant prose, to translate the wording of the optimistic realtors in which "charmingly secluded" means "you are snowed in for the winter" and "opportunity to remodel into your dream house" means "it is so rickety it will have to be rebuilt." At first we could not decide even on a general area. Soon, however, we realized that everything within the fifty-mile circle was too close to New York. It all seemed suburbs or suburbs-to-be. Even the 100-mile circle enclosed much that was fated to be overrun in the almost immediate future.

This was the way matters stood on a November evening when I attended a meeting of the Brooklyn Entomological Society. The speaker was Dr. James A. Slater, of the University of Connecticut. After the meeting, he suggested: "Why don't you try the northeastern corner of Connecticut? It's still largely rural. There is a beautiful little village up there named Hampton. We call it the Gem of the East."

Before the week was out, Nellie and I were driving down the single main street of Hampton, its white houses and spacious lawns and sugar maples occupying high ground at the edge of an abrupt escarpment that provides a magnificent vista stretching away to the east over the valley of Little River and the rolling hills beyond. We wandered on side roads through the companionable country around it. All this far northeastern corner of Connecticut is a region of little brooks, of villages, of hill-and-valley views. Even in November this was a fair land. The more we saw the more sure we were. This was it! This was what we had been looking for! We drove home that day in great relief. At least we had narrowed down our search. Now all we had to do was to find the right place—and find it for sale.

A short time later, back on Long Island, I received a letter from Wendell Davis, then the First Selectman of Hampton. He too was a professor at the University of Connecticut. The parents of his wife, Alison, had an early Cape Cod-type house for sale together with eighty acres of land, part of it wild. The photograph enclosed showed exactly the sort of house we had in mind. Luck seemed with us. We hurried back to Hampton. All was well—all except for one insuperable obstacle. The wide ribbon of concrete carrying the traffic of U.S. 6 cut the place in two. The house and a few acres, on one side of this major highway, were separated from the rest of the property on the other side. Cold rain and winter fog accentuated our disappointment as we rode home under heavy skies.

That winter Wendell and Alison kept track of all the Hampton places that were or might be for sale. We will always remember their months of kindly interest and helpfulness. Others, especially the Ostbys, Raymond and Leila, went out of their way to aid us. In fact, I discovered that, although we had come to Hampton without knowing anyone, a flattering number of people were familiar with my books and they all seemed anxious to assist us in our search. Toward the end of winter, the Davises sent us a list of half a dozen country homes that were on the market. As soon as the snow was gone in early April, our hopes revived, we were back in Hampton.

In the days that followed, those hopes receded again. Regretfully, one after the other, we crossed off the places on the list. In some the house was satisfactory but the land was not what we desired. In others the land was what we wanted but the house was in too bad repair or of the wrong design. What we were looking for, we decided, was a well-preserved house in run-down surroundings.

In desperation, for nearly a week, we rode with real estate men to see property for sale outside of Hampton—in Eastford and Ashford, beside the Quinebaug, near Snake Meadow, in Killingly, in Pomfret, and in Windham, the scene of the legendary Battle of the Frogs in pre-Revolutionary times. But always we circled back to Hampton. To find such a place as we had in mind, to find it for sale in this one small area—four by seven miles—seemed more and more impossible.

But later that month we were back again. The fact that we had agreed to sell our Long Island home to friends from Indiana, Howard and Henrietta Carey, increased the pressure on us to find our place in the country. All the houses on the block in which the Careys lived in Brooklyn were to be torn down for a housing development. They occupied the middle of the block and the destruction was approaching them from both directions. One telephone call would tell us that half the houses on the right were already gone, another that two more houses on the left had been demolished. Finally theirs was the only house left standing.

The evening of the twenty-ninth of April we spent with Wendell and Alison Davis, discussing all possibilities. Born of desperation, my suggestions grew wilder. Could I find a house we liked and move it to land that suited us? Could I rent somewhere in the neighborhood for a year while I looked for the right place to buy? Then Alison remembered that a few years before Margaret Marcus, a widow living on Kenyon Road, had talked of selling. Neither of the Davises knew much about the place except that the house was set well back from the road and that there was a brook that flowed somewhere nearby. Wendell put in a telephone call. While Alison, Nellie, and I sat on the edges of our chairs, hearing only one side of the conversation, he learned that while she was undecided whether she wanted to sell or not, she would be willing to have us look at the place in the morning.

So it came about that on the last day of April, about nine o'clock in the morning, on a quiet country road two miles north of the village, we turned off into a long brush-bordered lane. We advanced between stone walls and crossed a winding brook on a great slab of rock that for more than a century and a half had formed a solid bridge. Our first glimpse of the hidden house revealed a white cottage on a knoll beside a line of towering hickory trees—just such a cottage as we had pictured in our minds. We ascended the knoll while a green vista of meadowfields and woods expanded around us. Once again, as on that first day we visited Hampton, we knew a soaring sense of elation. This was it! This was what we had been looking for! We stopped at the back door—you almost always go into country houses through the kitchen.

Margaret Marcus, a widow caring for an ill aunt, was someone we liked from our first moment of meeting. Now, years later, time and further acquaintance with her kindly and responsible nature has added admiration to that liking. For several years, off and on, she had been considering selling. I think in the back of her mind she wanted to be sure only the right kind of people bought the place, those who would have the same affection for it she and her husband, Axel, had

had. On this point she was soon satisfied. Moreover when she had looked out the kitchen window as we were getting out of the car, she was amazed by Nellie's close resemblance to a special friend of hers she had not seen for years. So we all started off on a friendly footing.

When we entered the kitchen, as she was showing us through the house, we caught a glimpse of the great stone fireplace in the long living room. Going from room to room and from basement to attic, we saw pegged beams and floors of wide boards held in place with square hand-wrought nails. We saw the three fireplaces ranged around the central chimney. Cut into a small rectangular block of marble high on the outside of the chimney was the date when the house was built, 1806, six years before Napoleon retreated from Moscow, only thirty years after the signing of the Declaration of Independence. The old house, after weathering the storms of 153 years, was still solid and in good repair. Around it extended seventy-nine acres of varied land. It contained woods, open fields, swamps, two good-sized brooks, and a waterfall. In my mind I ticked off the list of things we had hoped for in a country home. Miraculously they all seemed here.

And after talking to her daughter, Peggy Fox, Mrs. Marcus decided she was ready to sell. Following a long attack of lumbago and troubles with an ailing furnace, she dreaded the idea of going through another winter. We had come, it turned out, just at the right time. Our bad luck was balanced with good luck in the end.

For so long, in the back of our minds, we had planned this move to a wilder, more primitive, more natural home that it was difficult for us to realize that it had come at last. That afternoon we all "walked the bounds" to see the woods and open fields of the land that would be ours.

CHAPTER TWO

Walking the Bounds

In a time when "pedestrian" is most often used to describe someone hit by an automobile, it was pleasant that afternoon to walk on mossy trails through the spring woods—sometimes beside a brook, sometimes along an old stone wall, sometimes skirting the edge of a swamp. Everywhere we went what we saw exceeded our expectations. For the least valuable land to others is often the most valuable land to the naturalist. It was the brush tangles, the shaded ravines, the wet lowland woods, the mink brooks and deer trails and redwing swamps, the wildest acres, that appealed to us most. Although Hampton, air line, is only 120 miles from New York City and only sixty miles from Boston, it lies in a region inhabited by deer and raccoon, fox and mink, muskrat, otter, and beaver.

Our walk started at the house, at an elevation of 526 feet above sea level. We descended the slope of the knoll toward the east, the same slope on which I later laid out an "insect garden"—a maze of interconnecting paths cut through weed tangles with a sundial at their center. Hardly more than 100 yards below the house and slightly upstream from the bridge and the lane we came to Hampton Brook flowing under its bordering ash and maple trees. Beyond, pastureland extended to Kenyon Road. In all directions, wherever we looked, woods and fields surrounded and protected the privacy of the secluded farmhouse.

We turned upstream. Cascading over rocks and pausing in shadowed pools, the brook flowed toward us bordered by lush lowland growth, by marsh marigold, skunk cabbage, and false hellebore. Behind us, the watercourse descended for another half mile to join Little River and meander on down the length of the town of Hampton.

As we ascended the stream our ears caught above the ripple of the current the sound of falling water. Where it emerges from the woods, the whole brook tumbles in a shining cascade over a dam of fieldstone. Originally the pool below this waterfall was used for washing sheep. In this region, from earliest times, sheep and cows, wool and mutton and milk, have been the chief sources of income from the stony fields. About forty acres of the farm consisted of open land. The rest had always been or had gone back to woodland.

Somewhere to the west of the waterfall, along the edge of a pasture, we plunged into the woods. Our path wandered north, first through an area that a few weeks later would be a dense jungle of cinnamon and interrupted ferns, then across the mossy rocks of a smaller stream, called Fern Brook, then on amid sassafras, hickory, oak, and maple. Once, during a later year, Nellie set out on an ambitious project—to count all the trees growing in our woods. A forester friend came to her rescue. He supplied the figure for the average number of trees to an acre in woods such as ours, counting everything as a tree that exceeded ten feet in height. Multiplying this figure by the number of our wooded acres, we discovered that the tree population of Trail Wood totals more than 22,500.

Below a hill covered with second-growth oaks, our path followed the edge of a swamp to emerge on the remnant of an ancient highway that Margaret Marcus always referred to as the Old Colonial Road. The ground was still packed hard from the wheels of wagons and carriages and, some say, stagecoaches that once traveled over this long-abandoned way. A little farther on, this track of a road ascended in a steep climbing turn up the flank of a ridge. Near the top it passed the silvered stub of a huge chestnut tree, a landmark in the area. Nearly half a century earlier the tree had died in the blight that swept through New England woods. Its shattered trunk towered like a monument engulfed by the woodland, a monument to the adaptable, invaluable wood and the sweet-flavored nuts of this tree now gone.

Later when we measured the stub with a steel tape, breast high, we found it had a circumference of fourteen feet. All through our woods we came upon the moldering stumps of giant chestnuts. Fast-growing, light, easy to work, and long-enduring, the wood of such trees was highly valued and is greatly missed. The sill under the front door of our farmhouse, placed there more than 160 years ago, is made of chestnut. It has never been replaced and it is still solid and undecayed.

We took our time, that afternoon, stopping to examine beautiful unfamiliar mosses, pausing at wide carpets of groundpine, halting beside slopes shaggy with Christmas ferns. Once along the northward-running ridgetop, we wandered among immense and weathered boulders that had lain there since the retreat of the last Ice Age. For fully half a mile the woods ran on in a narrowing wedge of land until it came to Griffin Road, where Cooney Bridge crossed the old right-of-way of what once had been the main railway line between Hartford and Boston. Past these woods, along this right-of-way, near the turn of the century, had run the famous high-speed passenger train, the White Ghost, with all its coaches painted white and its conductors clad in white uniforms.

It is this unused right-of-way that forms the western boundary of much of the farm. Looking across it, we saw the eastern edge of the Natchaug State Forest. Thus the land we were buying was separated only by the thin strip of the former railroad from the wild tract of one of the largest state forests in Connecticut. Old stone walls that we encountered running in haphazard fashion through the woods in this area dated back to before the building of the railroad. Farmers on opposite sides of the line, finding themselves cut off from part of their farms, traded land back and forth. In consequence, the original logic in the position of the walls was lost and now they appeared to wander aimlessly among the trees.

That western boundary, formed by the railroad, was clear-cut. But the eastern boundary of the land was less certain. On tussocks of grass we crossed a swampy stretch, following traces of an ancient barbed-wire fence, discovering here and there rusty fragments protruding from the trunks of trees that long ago engulfed the wires. For a time the boundary followed a tumbledown stone wall. Then it accompanied the meandering wheel tracks of a woods road over which, in earlier days, logs and firewood had been hauled. To straighten out this line and give us a firm boundary on all sides, I purchased, a few years later, additional land from our next door neighbor, Marian Pickles. Included in it were two swamps, a cranberry bog, and the full length of the old woods road. This purchase brought the area of the place to a total of—to use the phrase that appears in farm deeds—"more or less" 130 acres.

Returning southward by other paths, we walked the bounds of the lower woods. Through spicebush tangles, across the boggy ground that marks the beginning of Fern Brook, through a region of shagbark hickories and across another glacial boulderfield, we accompanied a long straight stone wall, gray and lichen-covered. It led us to the edge of New Hill Road, the southern boundary of the farm. Griffin Road, on the north, and New Hill Road, on the south, are nearly a mile apart.

In completing our circuit, we found ourselves on a wet woodland slope where the trickle of seven springs joined to form a little brook; we found ourselves at the highest point on the farm looking across the valley of Little River; we found ourselves on a dry hillside, a former pasture, covered from end to end with dark clumps of juniper. Almost from minute to minute, in this landscape sculptured by the glaciers, conditions changed around us—open woods, deep ravines, tangles of underbrush, dry ridges, glades, and wetlands. A visiting biologist who once followed our trails with us observed that he had never encountered a comparable area with so great a variety of habitats.

And everywhere we went we encountered a web of trails that spread across the pastures and the woods, trails laid down by human feet and by wild inhabitants of the area—deer trails, fox trails, rabbit trails, chipmunk trails. They all, for us, represented an important feature of the farm. Once, with a pedometer hooked to my belt, I followed the paths we walked that afternoon and others that Nellie and I opened up later on. The total length of these green pathways, in the woods alone, came to nearly three miles. Because of this interconnecting network of trails that has enabled us to observe from day to day all the changes of the successive seasons, we early decided on a name for our home in the country. The name we chose was Trail Wood.

Coming down from the high point of the farm, across the juniper-covered hillside and through a gap in the high sheep wall at its base, we followed tracks that deer had left in the night and saw where they had nipped off the tips of blackberry canes along the way. Close beside the sheep wall a small striped garter snake slipped into a tangle of weeds. Here there are no poisonous reptiles and, for a land so wild and rocky, snakes of any kind are extremely rare.

Now, between us and the white house that we could glimpse at times through the trees, there lay a bowl of land containing a red-maple swamp with its shallow pools of standing water and its soggy carpets of sphagnum moss. Around its western edge extended an alder swale. We pushed our way through it, stepping across a small stream that spread out and came together among the roots. To our right we could catch the glint of water shining amid the dark tree trunks. It came from a dozen or more large springs scattered across the basin. It was here, four years afterward, that we cut the trees and, partly by digging and partly by damming, formed a pond, an acre in extent, down the slope from the windows of the house.

We climbed that slope in the late afternoon. We had walked the bounds. Two days later, after legal formalities, we signed an agreement of purchase and on a Thursday in June, the eleventh day of the month, in 1959, we moved to our home in the country.

In the cool of the evening, at the end of that ninety-degree moving day, we sat on the terrace above the slope descending to Hampton

Brook and watched the silent winking lights of the fireflies and listened to the faraway calling of a whippoorwill. In a curiously vivid emotion, we felt that, in both the house and the farm, we had come home. Both had an old lived-in, almost familiar quality. I wondered how much of this emotion, for me, was due to the fact that Hampton lies on almost the same line of latitude as Lone Oak, my grandfather's dune-country farm in northern Indiana, that I knew so well in childhood.

Sitting there, that first evening, we took stock. Where were we? Hampton, the map shows, is about forty miles east of Hartford, forty miles north of New London, forty miles west of Providence, Rhode Island, and forty miles south of Worcester, Massachusetts. In the location of this small rural community we were at the axis of a wheel of cities. We were also in a land of picturesque names. Ten miles to the south of us there is a Toleration Road; six miles to the west of us a Carefree Lane. To the Indians, all this region was the Fresh Water Country. And nearby Windham, in early days, was known as the Hither Place.

Here we were also in a region rich in historical associations. Half a mile down Route 97 the Henry Beckers live in "the House the Women Built." In the days of the Revolution, when all the able-bodied men in the community had gone to war, a young bride, Sarah Hammond, determined to build a house and have it ready for her husband when he returned to Hampton. A one-legged carpenter, unable to join the army, prepared the framework. Then women of the village helped her complete the dwelling. Hardly seven miles from Trail Wood in the other direction, toward the north, lies the famous Wolf Den, the cavern among the rocks which General Israel Putnam, celebrated in both the French and Indian War and the Revolution, is said to have entered alone to kill the last renegade wolf in the region.

During Colonial times, Hampton lay between the country of the Mohegans and the Nipmucks. Just to the east of us an ancient Indian way, the Nipmuck Path, ran north to the Wet Flag Meadows and, beyond the Massachusetts line, to the lake with the longest place-name in America: Lake Chargoggagoggmanchaugagoggchaubunagungamaugg, near Webster. The name means a fishing place at the boundaries, a neutral meeting ground, or, as the more common translation has it: "You fish on your side, I'll fish on my side, and nobody fishes in the middle."

Long ago, on a New York street, I heard someone say to a companion: "I have always maintained that that was the opposite of a coincidence." At the time I wondered what the opposite of a coincidence would be. Our coming to Hampton, I suspect, could be so described. It was no accident, no happen-so. We worked for years to attain it. Very rarely, probably, do people move, as we did, to a place where they know no one, where they have no relations, where they are not sent by their work, where they have selected the locality entirely because it suits their special needs.

One of the attractions of Hampton for us was the fact that in 1959 its population was actually smaller than it had been in 1790. Our corner of Connecticut has been referred to by the government in Washington as an "industrially depressed area." The prediction of the Windham Regional Planning Agency is that Hampton will remain substantially rural through 1980. The inhabitants of the 25.3 square miles of the town now number fewer than 1,200, 300 of whom represent the village itself. When, in the 1960s, a scientist from the U.S. Geological Survey made a study of the ground water in the Hampton Quadrangle, he concluded it was insufficient ever to meet the demands of a large population.

A mile below the village, and 153 miles from its beginning at Provincetown on Cape Cod, U.S. 6, the longest highway under one designation in America, passes on its way to Long Beach on the coast of California. During our wandering travels through the American seasons we had encountered this cross-continental highway along most of its length. Each of those four journeys had been a great adventure. Now we were starting an adventure of another kind at Trail Wood. Canyons and deserts and rain forests and mountain tarns have the excitement of the strange, the out-of-the-ordinary. But in such a country place as ours the near-at-home also has its special attraction, its continuing charm, the hold of the familiar that is also the ever-changing.

Sitting there in the twilight, watching the fireflies and listening to the whippoorwill that first evening, we seemed in the perfect habitat for a pair of naturalists. We felt as comfortable as a rabbit in its form. Here, in every season of the year, we would be living on the edge of wildness. All these acres around us, all these fields and woods fading into the night, would form a sanctuary farm—a sanctuary for wildlife and a sanctuary for us.

CHAPTER THREE

The Long Lane

During the days of that first summer at Trail Wood we explored our new surroundings. Like expanding ripple rings on a pond, our investigations spread out from the house to the yard to the lane to the neighboring fields to the remoter portions of the woods and streams. Early, as an aid to making journal entries and reporting where we had been and what we had seen, Nellie and I began naming special features of the landscape. Before many weeks had passed, we were referring to Juniper Hill, Firefly Meadow, Groundpine Crossing, Monument Pasture, the Old Woods Road, and Pussywillow Corner at the end of the Long Lane.

In *Specimen Days* Walt Whitman wrote: "As every man has his hobby-liking, mine is for an old farm lane. . . ."

Our lane, a long lane, a wild lane, a real farm lane, extends east and west for nearly a tenth of a mile between the house and Kenyon Road. Stone walls parallel it. Enclosed between these walls and the wheeltracks run green bordering strips dense with bushes and ferns and wildflowers. Covering its length so many thousands of times, we have come to know the lane inch by inch. We have watched the changes that take place along it through all the seasons and from year to year.

When John Muir was living in his cabin beside the Merced River in Yosemite, a century ago, he became absorbed in studying the immense differences in the plants on opposite sides of his 3,000-foot-deep valley. The south side was shaded and cool, the north side warm and exposed to the sun. Similarly our lane, running east and west, has its cool southern border, shaded by the wall during much of the year, and its warmer, sunny northern side. One is the side of ferns, the other of wildflowers. One is mainly green, the other multicolored. One is the early side, the other the late side. The blooming season comes first to the plants of the sunlit northern border. There wildflowers grow in greater variety and greater profusion—rue anemone and wild columbine, the fringed loosestrife and showy tick trefoil and meadowsweet, large-leaved asters and the beautiful wood betony with its flowerheads two-toned, yellow and red.

At first glance the predominantly green southern side seems less interesting. But a second glance is more revealing. One summer day, between the house and the road, we counted fourteen different species of ferns, more than one-fourth the number recorded for the whole state of Connecticut. In my mind's eye I can see them still: the soft and feathery light-green masses of the hay-scented ferns; the royal and sensitive ferns and the silvery spleen-wort rooted in the moister soil near the brook; the dark rich green of the evergreen wood or marginal shield fern, a plant that retains its color all through the subzero days of winter; the wall of interrupted ferns rising, rank on rank, as high as my shoulders; the low ternate grape fern; the lady fern; the Christmas fern with fronds that, like those of the evergreen wood fern, remain green the year around; the shade-loving spinulose wood fern, marsh fern, and crested shield fern; and, near the entrance on Kenyon Road, the slender double-tapered fronds of the New York fern and the green fountains of the cinnamon ferns rising along the edge of the damp land of Pussywillow Corner.

For all of Trail Wood our census of native ferns has yielded twenty-six species, exactly one-half the fifty-two on the list for Connecticut. In addition to the fourteen along the lane we have found the cut-leaved grape fern, the leather-leaved grape fern, the rattlesnake grape fern, the polypody, the bracken, the fragile fern, the Boott's fern, the long beech fern, the broad beech fern, the maidenhair fern, the American shield or intermediate fern, and the ebony spleenwort. At various times friends have given us other species which have flourished where we set them out: the ostrich fern, the bladder fern, Goldie's fern, the net-veined chain fern, Braun's holly fern, and the narrow-leaved spleenwort. Adding these to the twenty-six, we now have thirty-two different species of ferns growing at Trail Wood.

A long hunt preceded the discovery of our first ebony spleenwort. We came upon it at last, a single plant growing in the boulder field along the ridge of the North Woods. Then, a few years later, we found that a westward-facing embankment on Kenyon Road was covered with these delicate ferns. I counted their fronds. The total came to an incredible 2,972, all growing in a space less than 200 feet long. During a subsequent summer, while we were away, a highway crew drenched the sides of Kenyon Road with herbicide spray. In one afternoon they virtually wiped out this spectacular stand of this beautiful fern. Gradually, little by little, it has been creeping back.

Oftentimes, after rain in the night, all the ferns along the south side of our lane are clothed in special beauty, their topmost fronds spangled with glistening drops of water. Every few steps I stop to peer closely at them. Each is crystal clear. Each is rounded on top, flat on the bottom. Each forms a miniature magnifying glass, a liquid lens. When I look through one, I see the tissues of the fern below magnified within the space of the small disk where the raindrop and the fern are in contact.

At other seasons of the year we follow the lane with the rays of the morning sun fanning downward in slanting lines through the autumn mist or with all the trees around us clad in a glittering armor of ice. Each season fills the air along our path with its own particular scents. Midsummer brings the intensely sweet fragrance of the tiny white flowers of the bedstraw; autumn the rich perfume of the ripe fox grapes; winter the smell of snow in the air, of hickory smoke on the wind; spring the first faint scent of a distant skunk amid the melting snow, then the primeval odor of wet fresh-turned earth, then the perfume of the returning wildflowers. In these daily walks along the lane, we balance the optimism of spring with the ripe experience of autumn. The possibilities of all things, the seeds of everything, appear inherent in the spring. But autumn is more mature, more reflective. It speaks of a time of recession as well as a time of advance, of a tide that ebbs as well as flows, of the stalemate of nature and the balance of the seasons.

Little areas all along our way hold special interest. Many are connected with the wildflowers of the lane. When bloodroot blooms beside the wall, I occasionally sit on the loam among the plants in the warm spring sunshine and examine through a little magnifying glass the waxy texture of the white petals of uneven length, the shining gold of the central stamens, the rich orange-colored sap that oozes from a broken stem, the pale-green veining of the large lobed leaves embracing the stalk. Sitting there, enjoying in magnified detail the beauty of those spring flowers, I bask in the sun like an outspread leaf. At such times I make, no doubt, my own equivalent of chlorophyll.

Blindfolded or groping my way along the lane in the dark of the moon, I could recognize when another flower blooms a dozen paces or so beyond the bloodroot. It fills the air around it with its own particular perfume—the overpowering dead-mouse smell of decaying flesh. Its scientific name is *Smilax herbacea*, but its eminently appropriate common name is carrion flower. In rounded sprays of small greenish-tinged flowers, the blooms appear on a vine that is related to the cat briar. Small flies and beetles, I notice, are attracted to them.

Several times, across the lane, among the tick trefoil, I have come upon a brilliant metallic-green little wasp anchored to a stem. It was lost to the world, fast asleep. Whenever we discover a new insect, our first question is: "*What* is it?" The answer in this instance is that it is one of the *Chrysididae*. But always there is a second question, or rather the same question with a different emphasis: "What *is* it?" And that is more difficult to answer. It encompasses the abilities, the habits, the life story of the individual. That answer usually entails patient and prolonged study in the field. In the case of the sleeping green wasp, observers have found that its life history parallels that of the cuckoo among birds. In a similar stealthy way it deposits its eggs in the nests of other hymenoptera.

Toward the end of summer, in the cool of dawn, all along the lane I find other insects anchored to plants where they have slept during the night. The most numerous are dark little bumblebees; the most memorable are small syrphid flies. I always find the latter clinging to the pendant flowers of the jewelweed, or touch-me-not. Often I will see a score or more within the space of a few square yards. Both the spotted orange blooms and the slender, large-headed flies are bedecked with innumerable glittering jewels that are tiny shining droplets of dew.

Where this stand of jewelweed grows it rises from ground carpeted with that remarkable vine that thrives in swamps, among sand dunes, in arid fields, in open woods, and by the edge of the sea, that plant that is such a menace to those allergic to its oil—poison ivy. Birds apparently are immune. More than sixty species consume the frosty-gray berries as autumn food. These berries, at first glance, closely resemble the fruit of bayberries. This fact was discovered the hard way, one fall, by a second-grade class in the Hampton school when a well-remembered substitute teacher from Hartford used poison ivy berries to demonstrate how to boil out wax for making bayberry candles.

Nature, along this portion of our lane, seems to have provided a ready antidote for the fiery itching produced by poison ivy. As the Indians did before me, I crush up the succulent stems of the jewelweed and rub on the juice to allay the misery. This juxtaposition of the two plants recalls a humorous instance of the way something is sometimes heard in youth and is repeated throughout a lifetime without one's ever observing whether it is true or false. One visitor to Trail Wood assured me in an authoritative voice that it is a well-known fact that jewelweed and poison ivy are never found together, that jewelweed prevents the poison ivy from growing near it. This he had heard as a child. For forty years he had repeated the assertion unchallenged. Yet that very day, in coming down the Trail Wood lane, he had passed the ivy and the jewelweed growing side by side.

Similarly all through the Dark Ages men repeated the erroneous statements of Pliny and Aristotle without ever subjecting them to the test of personal observation. Yet in every age there must have been some perhaps simple people who had never read the Ancients, who saw clearly the things occurring around them, who were aware of what was actually taking place. Even in Pliny's time there must have been some people who did not believe that bees carry pebbles to ballast themselves in the wind. Even in Biblical times there must have been some who doubted that honeybees are engendered in the decaying flesh of dead oxen. Even in Elizabethan years there must have been some who viewed with skepticism the belief that bats feed on bacon. But who were they? Nobody knows. They are forgotten. More properly they were never known. They were overwhelmed by the learned ignorance, the accepted say-so of their times.

Wet Weather Brook—an intermittent streamlet that comes to life in spring and in rainy weather and ceases to flow in summer droughts—in draining Pussywillow Corner passes under our lane through a rude culvert made of stones. Beside it and close to the southern wall rises a large and ancient white mulberry tree. It is one of three we have on the farm. They constitute a kind of monument to an insect that once flourished in this region but that disappeared generations ago. As early as 1760, before the Revolution, half an ounce of mulberry seeds was distributed to each parish in what is now Connecticut to encourage the rearing of silkworms. In the attics of hundreds of early farmhouses leaves from the mulberry trees were fed to the silk-producing caterpillars. The old Lyman Baker place just to the north of us on Kenyon Road was at one time known as the Mulberry Farm. The first silk mill in America—now at Henry Ford's museum at Dearborn, Michigan—was erected on the banks of a small stream at Gurleyville, only ten miles across the hills from Trail Wood.

In winter cottontail rabbits find shelter in the stone culvert under the lane. We see their tracks radiating out from the entrance in the morning. And foxes that hunt the rabbits come and go through a squarish opening, a foot and a half or so across, that pierces the stone wall near the mulberry tree. We refer to it as the Fox Door. Many other tracks besides those of fox and rabbit catch our eye along the lane. One summer day in the dust I came upon a wavering line of eight cuplike depressions. A large dog had left its pawprints and sparrows had settled in them and had enlarged them in their dusting. But the tracks that I remember most vividly were made not in snow or dust but in the soft wet soil after a night of summer rain. All the way to Kenyon Road I followed them—delicate clear-cut little hoofprints where a small fawn had walked that way before me.

A few pages back I commented on the little areas along the lane holding special interest. Of them all the outstanding one is the space around the old stone bridge that crosses Hampton Brook. It is always a stopping place on our trips to and fro along the lane. In warm spring rains we stand there listening to the drops drumming on the large leaves of the false hellebore. In the sweltering heat of August afternoons we watch damselflies with sable wings and slender bodies shining with metallic greens and blues as they make their way in slow fluttering flight over the water. Or on some early-autumn day we follow the wanderings of a jumping spider among the stones. I remember a hot summer morning when I was watching half a dozen small American copper butterflies whirling in the sunshine. There was a rush and the snap of a bill. A scarlet tanager zoomed back up to a telephone wire, carrying one of the butterflies. I saw it throw back its head and swallow the winged insect in a gulp as a heron swallows a fish.

About the time that June ends and July begins the air around the bridge is always heavy with the perfume of the white flower clusters on the elderberry bushes. August replaces the flowers with the masses of dark purple fruit. Toward the end of that month, one summer, as I came back along the lane about noon, I heard crackling sounds and saw the top of one of the elderberry bushes waving as though in a wind. Stealthily, I peered over the downstream wall of the bridge. A half-grown woodchuck, intent on reaching the juicy clusters at the ends of the twigs, was climbing the bush. It floundered among the slender branches. One broke. It lost its hold, turned upside-down, righted itself and began scrambling slowly upward again. Apparently it never saw me. Each time it looked in my direction it was blinded by the sun. It had hitched itself about halfway to the top of the bush when it gave up. Awkwardly it turned around and headed down. It slid the last few feet to the ground. Then it looked back into the bush, paused to pull down and nibble off a raspberry leaf, and disappeared into the weeds.

This particular woodchuck I knew well. It had been born in a burrow near the hay-scented ferns beside the wall. In the days when it first appeared at the entrance of its burrow it used to make snuffling sounds and whistle at me as I went by on my way for the mail. On days when it was not in evidence, I found I could bring it to the opening by imitating its shrill call. Sometimes we used to continue for a minute or two calling back and forth, giving whistle for whistle, before I moved away.

Besides its varied natural history interest, our lane, of course, has its practical value. Down it come our mail, our newspapers, our magazines. Along it run the wires of our electric power and our telephone. It provides a road for the delivery of milk and laundry and fuel oil. It also has its human side. For along it have come visitors to Trail Wood.

The first to turn into the lane, a few days after we had moved, were Roger Tory Peterson and Barbara. They drove up from Old Lyme to see us in our new habitat. Since then our lane has brought us a diversity of interesting people—a schoolteacher from New Brunswick, another from California, an editor from London, England, a museum head from a Western state, a scientist who had traveled to the Orient to study the synchronous flashing of fireflies. Past the mulberry tree and over the bridge and up the knoll have come Leslie Peltier, the famous amateur astronomer and discoverer of comets, Samuel Gottscho, that remarkable ninety-year-old photographer of landscapes and wildflowers, and Farida Wiley, authority on our native ferns. One young couple from Saskatchewan, Bill and Vicky McMillan, left behind a mobile formed of cutouts of the leaves of the trees of Saskatoon. Another,

Toby and Janet Schuh, had just returned from a year spent in the jungles at the headwaters of the Amazon.

When a couple arrived from Boston, bringing some of my books to be autographed, our conversation never did progress very far. They kept interrupting to exclaim over and over: "Oh, the peace and the quiet!" During a later summer, a Midwestern couple, Mr. and Mrs. Carl Scheffler, drove all the way from Ypsilanti, Michigan, to see us, then turned around and drove back again. We had never met. But they had read all my books and felt they knew us both of old. When Nellie appeared they exclaimed:

"Why, there's Nellie!"

Another visitor, from Fort Worth, Texas, had never seen any of my books or heard of us. But as a small boy he had lived for a year or two in the country near Hampton. He had always remembered this early home with affection and, while on a trip north, he had come back and was going from farm to farm hoping to locate the place once more. He handed me his card. He was a member of the Texas legislature, Winthrop C. "Bud" Sherman. All he could remember about the farm where he had lived so long before was the general lay of the land and that there was a waterfall and a brook that flowed near the house. The more he wandered about Trail Wood the more sure he was that he had found the place for which he was searching. So delighted was he to discover it again that each year since at Christmas time he has sent us a box of giant Texas grapefruit—one of the many fringe benefits that have come to us from our original good luck in finding our sought-for home in the country at Trail Wood.

CHAPTER FOUR

A Hammock in the Woods

One summer day, a woman from the city, lost in the country, drove up our lane to inquire directions. She sat in her car looking around. Chipmunks raced along the stone walls. Birds sang in the trees. A woodchuck sat up in a pasture.

"What kind of a farm is this?" she asked. Then after a long pause she said: "Oh, I get it. It's a fun thing."

That is not exactly the way Nellie and I think of Trail Wood. For us it is a farm with a different kind of harvest. We are farmers who cultivate a different sort of crops. Our fields are unplanted. But they are not unused. The yield for us is made up of observations and memories, of greater understanding and little adventures by the way. But, I must admit, a strong element of having fun has run through all our days at Trail Wood. Take, for example, my hammock in the woods.

During our second summer I bought a heavy canvas hammock at an army and navy store. With its stout ropes I could tie it between trees beside the brook or sling it beneath some strong lower limb in the woods. There it made a superb observation post. For, after it had been in place a day or two, the wildlife became accustomed to its presence and failed to notice any difference when I was lying in it. Thus during summer days along the Old Woods Road, in Wild Apple Glade or close to the waterfall on Hampton Brook, I have been able to experience rare hours observing the life around me.

Usually, along with my binoculars and pocket notebook, I took some small volume to read during lulls in the activity. Under such circumstances, I remember, I read Poe's tales, Pascal's thoughts, Ossian's poems, and Leonardo da Vinci's notebooks. But always the special attraction was the opportunity of seeing wild creatures going about their daily affairs, unaware that human eyes followed them in their comings and goings among the trees.

Charles Darwin noted that in our everyday life we rarely look higher than fifteen degrees above the horizon. In my hammock my view began where it ordinarily ends. My eyes were directed almost entirely upward. They ascended the bark of tree trunks and strayed from layer to layer among the leafy boughs. They swept through the air with the flight of birds and leaped from tree to tree and coursed along the branches with the squirrels.

Over the years, a dozen times or more, in woods and forests on days of perfect stillness I have been startled by the sharp crack of a breaking branch. With no wind or other apparent cause, somewhere around me a limb had fallen to the ground. Only once have I seen such a dry and brittle branch at the very moment of its breaking. And on that occasion I was lying in my hammock not far from the Old Woods Road. A small flock of half a dozen bluejays, uttering short call-notes reminiscent of the tuning of violins, was moving among the swamp maples. One, as I looked up, alighted high in one of the maple trees. Just as its feet gripped a dead branch, it snapped off with a quick, explosive report. The limb dropped to the ground; the bluejay, in a flutter of wings, ascended to a higher perch. Many a bird and many a squirrel, no doubt, provides the slight final strain that breaks some long-weakening branch among the woodland trees.

On the hottest days of summer, lying in the shade beside the brook, it was pleasant to listen to all the natural sounds around me and in imagination to trace them to their sources. Catbirds mewed and towhees called their clear "che-winks" and, under the cumulus clouds high overhead, a soaring red-shouldered hawk repeated its scream, that shrill, wild, far-carrying "kee-you" the bluejays so frequently imitate. Flickers sometimes alighted on tree trunks close by, uttering a soft fluttering sound that brought to mind the call of a gray tree frog. In the woods the red-eyed vireo went on and on, unwinding a song that advanced like the uneven descent of rapids, the phrases separated by short pauses, bird music far different from the smooth flow of the thrushes.

Where wet moss cushioned the rocks beside the brook, green frogs occasionally gave voice to sounds that at times suggested the low-pitched twang of a banjo string, at other times two rocks being grated together. After long periods of silence all the frogs would begin to call. What set them off? Nothing I could sense; but something, without doubt, the frogs sensed well.

Always during the warmest part of the day the high-pitched soaring song of the cicadas rose and fell among the treetops. Once one of these insects, passing from tree to tree, flew low over my hammock, its call sputtering on and off like a defective engine. Perhaps such intermittent sound helps protect it while in flight, making it more difficult for enemies to pinpoint its exact location in the air. On such days the breeze blew, the leaves rustled, the brook murmured, the frogs strummed, the vireo sang, the descending call of the cicadas sizzled to an end only to begin again—so the drowsy minutes passed.

During the latter days of spring, under the brookside trees, I often saw young birds, whole families just off the nest, go trooping overhead, fluttering and pausing, advancing from branch to branch and treetop to treetop. One morning in particular returns to mind. Lying in my bird-watching hammock, I looked up while group after group paraded by. I cannot remember another time when I witnessed so many species of birds breaking nest ties in a comparable area and period of time.

First came a family of grackles with the harsh sound of their complaining voices. These dusky-hued birds were soon succeeded by two colorful groups arriving in succession, northern orioles followed by scarlet tanagers. Continually the vivid plumage of the adults flashed in and out among the green foliage above me. Next to appear were the phoebes. As the two parent birds hawked about for fluttering insects, swooping close around my hammock, the snapping sound of their bills as they made their aerial captures came to my ears like miniature explosions. The final group to move above me that day were the kingbirds. Excitement accompanied them as they advanced. Their strident challenging calls filled the air. And when two crows came skulking down the line of trees, fireworks erupted. In a screaming attack and pursuit, one of the adult kingbirds plunged down and, clinging to the back of a crow for several seconds, rode it through the air, whacking mightily with its bill.

As might have been expected, crows were usually the birds that first sighted me lying motionless in my hammock. One afternoon, however, I saw two of these ebony birds alight uncertainly among the branches directly overhead. They were completely oblivious to my presence. When they cawed, their higher, almost falsetto voices proclaimed them young birds recently off the nest. A minute or two later an adult came winging along the brook, lower down. It spotted me, veered away, cawed violently in alarm. Instantly the two birds above me floundered among the upper twigs as they struggled to become airborne. All three flapped away, veering on a zigzag course that kept the denser foliage of the trees between them and me.

On occasions, as I lay in my hammock, I pulled gently on a small rope or a double strand of binding twine I had tied to a branch to one side and so kept up a slow rhythmical swinging as I read or looked about me. Toward the end of one sweltering day in July, as I was thus occupied, my eyes wandered along the weathered gray stones of a century-old wall about fifty feet away. They came to rest on something brown. With its flattened head placed on its forepaws on top of the wall, a woodchuck was watching me intently. It seemed fascinated by the regular movements of the hammock. For nearly fifteen minutes it remained perfectly still, its gaze unshifting. Only when I threw my legs over the side of the hammock and stood up did its scrutiny abruptly end as it disappeared in a scrambling rush behind the wall.

Of all the places where I enjoyed this relaxed form of nature observation, my favorite was Wild Apple Glade. At the center of this small secluded opening in the woods—like its hub—there rose the massive trunk of an ancient wild apple tree. Its gnarled limbs, many now barkless, spread outward in all directions. Its twisted roots writhed among mossy stones over the floor of the little clearing. One horizontal lower limb, as hard as metal, provided a solid support for my hammock and for me. Sometimes a woodcock would rise on whistling wings as I approached the glade and once, toward the end of the day, I startled a deer getting the small sour apples and saw it, its white flag shining, bounding away into the darkening woods.

This secret spot, this wild and lonely sanctuary, is hardly five minutes' walk from the house—around the end of the pond, up a slope among sweetfern and juniper, along an old stone wall that winds through the woods to the west. This path traverses a stand of blackberries and, a little later, skirts the remnants of the stone foundation of a long-gone hay barn. At one period in its past Trail Wood was known as the Blackberry Farm, and at another as the Hay Farm. Often on late-summer days I arrived at my hammock with a handful of red thorn apples or of blackberries filled with sun-ripened sweetness. These I ate at leisure as I lay in the stillness of the glade.

Once as I nibbled on such wild fruit gathered along the way, a movement caught a corner of my eye. A small covey of quail had emerged from among the surrounding trees. I froze into immobility, a thorn apple halfway to my mouth. With their soft "Toy! Toy!" calls, the birds advanced in little running fits and starts as they hunted for food along the woodland floor. Without realizing I was there, they passed directly beneath my hammock. I watched them as they crossed the glade, reached the opposite side of the opening and disappeared. Another time I was startled by the sudden muffled roar of a grouse's wings as the bird launched itself into the air almost beneath me. I had not been aware of its presence. But some small movement I had made had caught its sharp eye. It saw in me a stranger, a menace, a man.

In all probability, the limbs of the old apple tree had endured wind and rain, cold and drought, for nearly a century. Many had patches of gray-brown bark curling from the wood or hanging down. Looking up among these branches on a September day, I watched a mourning cloak butterfly flutter from limb to limb. It would alight, walk about, then take wing again. The resting place it finally chose was where half a dozen fragments of bark hung down below a dead or dying branch. There, anchored in place, it closed its wings and remained immobile. When I glanced away and looked back again, I had difficulty finding it. Its colors and form made it almost indistinguishable in the midst of the surrounding fragments of bark. Among all the limbs, it had chosen the spot where it seemed most perfectly camouflaged.

On certain days in summer, most frequently in the heat of afternoon, I would see butterflies of varied colors loitering about this woodland glade. I recall one occasion when, as I lay back, letting my eyes explore among the maze of the old apple boughs above me, I caught sight of a resting mourning cloak, then two blue-eyed graylings and a pearly eye and then a little wood satyr. The longer I looked, as I lay there that afternoon, the more of these airy insects I made out. The count rose to twelve, then to fifteen. From the lowest to the highest branches, the insects had found temporary resting places. For a time my hammock seemed slung beneath a tree of butterflies.

A great stillness settled over the woods on the hottest afternoons. I was lying in my hammock, half dozing in the heat of one such day, when I became aware of an ovenbird peering down at me from a branch just overhead. It stared in silence, round-eyed. Then it hopped from limb to limb, seeking a better view. Its prominent eye-rings gave it the appearance of watching me through white-rimmed spectacles. On another afternoon some movement of mine caught the attention of a wood thrush. I saw it staring intently at me from a tangle of wild grape vines that burdened a small oak at the edge of the opening, filling the glade with the fragrance of ripened fruit when September came. It was not until the thrush had flown from one tree to another and had made almost a complete circuit of the open space, pausing on each successive perch to examine me from a new angle, that its curiosity abated and it flew away.

Toward the end of summer the woods grew lighter, more yellow-green in hue. Gazing up, I would see some leaf let go and come drifting down toward me. In listening to the small woodland sounds on a still autumn day, I could catch the dry scrape and whisper of foliage fluttering down from all the trees around me. Now the hopping of birds and the progress of small mammals were reported in the crackle of the dry carpet that covered the glade. Thus I was able to follow each successive leap when a chipmunk, one day, entered the opening, passed beneath my hammock, and continued toward the other side where the little quail had vanished. In shifting my weight slightly to get a better view, I alarmed it. In short squeaking dashes, it raced noisily away over the fallen leaves. Then, from the comparative safety of a stone wall, it commenced a steady staccato chipping.

It is in the late summer and early fall that the chipping of these animals rises to a crescendo of repetition in our woods. At such times the chipmunk population is at its peak, families have separated, territories are being established. The loud "Chip! Chip! Chip!" is repeated over and over. It forms a territorial call, a warning to others of their kind, an auditory "No Trespassing" sign. It fills the woods during all the sunny days of early autumn. During these days, on mossy rocks, on moldering logs, on stumps and old stone walls, the calling goes on and on.

Soon after we arrived at Trail Wood I commenced counting the number of consecutive calls given by a chipmunk without pausing. The first year's record was 123. The next year that initial record was broken from the top of a mossy stump in the North Woods. The caller there chipped 348 times. So far the all-time record is held by a chipmunk that inhabited a lichen-covered stone wall beyond Hampton Brook along the Old Woods Road. Perched on top of this wall on an October day, it called 536 times before it fell silent only to begin again after a pause of no more than a minute or two.

During one of the early summers I waterproofed the canvas of my hammock to make it more resistant to weathering. After the first heavy rain when I came to Wild Apple Glade I discovered the cloth, now watertight, resembled an elongated trough half full of water. I dumped it out and let the hammock dry. Afterward, each time I left it, I folded over the sides—in effect rolling up the hammock—to keep out the rain.

This led to new complications. On the last day of August that year I folded back one side and uncovered a delicate oval nest woven of light-brown fibers. It resembled a cocoon with a round opening at one end. The structure appeared as fragile as a bubble. Through the almost transparent wall on my side, I glimpsed the eye of a terrified mother white-footed mouse huddling over her litter of young. She had woven the tan shell from frayed fibers pulled from the ends of the ropes that supported the hammock. A crevice in the canvas nearby held a cluster of the seeds of spicebush and wild cherry, each seed with an opening gnawed in one side.

As I bent over the nest, I saw the courageous little animal keep reaching up with its delicate forepaws, rearranging the fibers, pulling in the sides of the doorway. In a minute or two she had closed the opening entirely and had substantially increased the denseness of the wall through which I peered. Before I replaced the canvas, leaving this family of woodland mice hidden within its folds, I noticed that a tiny feather of a brownish bird had been woven among the fibers of the nest.

Fortunately baby mice mature rapidly. A week later the mice in their penthouse nest were gone and I was enjoying my hammock once more. But from time to time, during successive summers, other mice spent their earliest days swinging in my hammock. Each time my interest in their adventures more than made up for the inconvenience their presence caused.

One litter actually rode out the tail end of a hurricane protected from wind and deluge within the folds of canvas. When I came to the glade after the great storm was over, I found it littered with leaves and bark and broken twigs. All through the woods limbs and trees, particularly swamp maples, had gone down. But the old wild apple tree, the hub of the glade, was still intact. The storm was but another incident in its long storm-filled life. When I opened the hammock and took stock of the interior I found all the white-footed mice snug within. They had been tossed about on the windy seas. They had swung through the air with all the gyrations of their supporting canvas. But through it all, until calm returned again, they had remained sheltered and unharmed. This one and only life—for mice and men—how strange and uncertain and adventurous it is!

An entirely different time of peril befell another litter of hammock mice on a still and sunny morning in mid-September. In trying to take a closeup photograph of the nest, I accidentally jarred it in a sudden jolt. The frightened mother mouse took flight with all her young. Pushing her way out through the end of the nest, she appeared dragging her whole litter behind her—four baby mice, about half grown, all clinging desperately to her rear quarters. They slid behind her like a quartet of small gray sleds as she ran up the slope of the canvas. For only an instant did she hesitate at the edge of the hammock. Then, with the litter still clinging in place, she launched herself into space.

Later I measured the distance through the air from the hammock edge to the spot amid moss-covered roots and stones where the cluster of mice struck the ground. It was thirty-nine inches. The light weight of the mice stood them in good stead. The cushioning effect of the moss also aided. They struck and bounced. But they landed without breaking any bones, apparently without even having the breath knocked out of them, and with none of the baby mice even losing its grip on its mother's body.

She appeared to hit the ground running. With the babies bouncing along behind her, she scrambled over roots and rocks. She reached the trunk of the wild apple tree, started to climb it, then changed her course, ran first one way, then another, desperately seeking some sanctuary on the open floor of the glade. The whole cluster of little animals became wet with dew. Frequently I saw the mother pause and rub her forepaws rapidly over her face and vigorously clean the fur of her breast. This seemed done partly to remove the dew and partly to relieve nervous tension. She repeated the performance every minute or two. At such times she appeared wringing her hands in anxiety. I wished I could tell her I posed no threat. But the only language I could use was the language of action. So I drew back among the trees at the farther side of the opening. And here I saw her reach a hiding place at last, a spot where ferns hung low, overarching a tangle of twisted tree roots. In saving her family by these desperate exertions, this small animal, so beautiful and so resolute, has left an enduring memory.

In the course of time, as other summers brought other families of woodland mice, nests with white fibers among the brown appeared. The material came less and less from the raveled ends of the ropes. The mice had hit on a better idea. They had found a source of material nearer home by simply gnawing holes through the canvas. My hammock, as new holes appeared, grew more and more to resemble a Swiss cheese.

But it had served its pleasant purpose. At summer's end, one year, I untied the ropes, rolled up my hammock for a final time, and carried it home on my shoulder. There were other diversions to turn to. Particularly there was the fun of a hollow brushpile, another adventure in viewpoint that will be recounted in a later chapter.

CHAPTER FIVE

Wild Meadows

"In Upland mowings now no longer mown/The banished weed again lifts high its head,/Ennobled by some quaint ancestral name." So Benjamin T. Richards wrote of our Trail Wood meadows in a sonnet, "The Naturalist Buys an Old Farm," published in *The New York Times* the year we moved to Hampton.

Green in the spring and golden in the fall, these fallow fields reclaimed by flowers, wild meadows all, spread away from our house on the knoll. In every direction they lead our eyes over rolling open land to the background of the woods. Looking across these fields, we watch the same seasonal sequence year after year—the blooming of the same wildflowers, the flocks of gray-clad young starlings pouring up from their feeding in the August grass, the southward drift of the monarch butterflies, the glistening threads of gossamer where ballooning spiderlings have come to earth.

Each year, as summer ends and the first waves of migration begin, our pastures overnight are sown with robins. They also become swallowfields at sunset with hundreds of white-breasted tree swallows sweeping low back and forth over the grasstops. They become flickerfields, where the flash of white rump patches reveal where the migrating woodpeckers are alighting and taking off in their search for ants.

During our first few summers at Trail Wood, black-and-white Holstein cows grazed in our pastures. Clarence and Walter Stone brought them down from their dairy farm on Grow Hill to the north. All had been bred and were awaiting their calves. Once on the same day, a Friday in June, three calves were born in the space of a few hours. On another day, as Nellie and I were starting out to walk along the trails of the North Woods, we passed a particularly white Holstein cow placidly chewing her cud by the bars. When we returned, a little more than an hour later, a black-and-white calf was running about without a wobble. By the time a calf is an hour old it can often outrun a man.

The wildest of all the calves born in our pastures arrived on a morning in July. A white triangle on its forehead distinguished it. For days afterward it appeared and disappeared. It slept hidden like a fawn. When at last it was cornered, it took both the Stone brothers to hold it down. All the calves produced by its mother possessed an untamable streak. One remained uncaught for six weeks. It outran and outdodged seven men who sought to close in on it in a corner of a field. Probably no bison calf on the prairie was ever more wild than that six-week-old animal in a New England pasture.

Watching the cows became almost as interesting, for us, as watching wildlife. We saw one white-faced heifer, called Snowball, experiment endlessly with new foods. She browsed on trees like a deer. She consumed even the leaves of spicebush and prickly ash. And her consumption of grass was varied with such unlikely items as catnip, mullein, pokeweed, and nettles. Another heifer seemed especially attracted to wildflowers. Where she had fed we found the tops gone from daisies and fleabane and narrow-leaved mountain mint, even from the stalks of cardinal flowers along the brook.

For a cow, eating is its main business. Oftentimes the summer days were not long enough for their feeding. More than once, in the middle of the night, we have glimpsed them methodically cropping grass in the moonlight. I can recall no instance of a cow in our pastures looking up into the sky. Most of a cow's life is spent with its head close to the ground. Because of this, one famous photographer of prize dairy cattle used to carry an umbrella on every assignment. When he had focused sharply on the animal he was photographing, he would open the umbrella and raise it at arm's length above his head. The cow would look up. While its head was lifted and its expression alert, he snapped his picture.

We were sometimes amused by the varied ways in which these animals revealed they were creatures of habit. Some inner timetable, inherited from the herding instincts of long-ago wild ancestors, seemed to regulate their movements. They all shifted their feeding from one field to another at about the same time day after day. They all traveled single file to the brook to drink together. One special tree trunk, where they crossed the stream into Monument Pasture, was used as a rubbing post. In aerial photographs taken in 1951 the central elevation of Monument Pasture has the appearance of a circular piece of corduroy. Parallel lines curve around the slopes of the hill—the cowpaths left by the feet of generations of cattle, each following the track left by those that went before.

At various times, after the cows had gone, our meadows were inhabited by horses belonging to neighbors—once by a beautiful pair of cream-colored Palominos, at other times by an Arabian horse and a Welsh pony and even, off and on during two summers, by a racehorse named Mighty Alf. He was the most interesting of all. After years on Eastern tracks, after winning races and setting records, a bowed tendon in a foreleg had brought his career to an end. With ears up and head held high, he would stand by the bars and wait for me to feed him a handful of fallen apples. Once when I clapped my hands to scare away a roving cat that was stalking a bird, Mighty Alf came running from a far corner of the pasture. I had accidentally discovered the signal that had been used to call him for a feeding of grain.

As he munched apple after apple, taking each carefully from my hand, I often tried to imagine myself inside his brain. Almost all his days had been spent amid noise and crowds. He had been surrounded by people, grooms, and trainers. He had moved constantly from place to place. He had been guarded, pampered, worked on schedule from his earliest years. Then, almost overnight, he had been plunged from this abnormally ordered and active life into an unending vacation, a time when he had nothing to do. No longer were there crowds, no longer excitement. During his first weeks at Trail Wood he appeared bored and apathetic. It was probably not the strangeness of a new place that bothered him—a racehorse lives anywhere, is used to travel and change. It was the sameness of the surroundings, the lack of variety, the absence of competition. Day followed day with predictable monotony in this almost lonely pasture. All during his first summer, we watched with fascination the silent drama of his gradual adaptation to a new and calmer life.

Then the horses went and our wild meadows became wilder. More frequently, along the remoter edges, we found rounded areas of flattened grass where deer had come from the woods and rested at night. New wildflowers appeared. The grass grew longer. The islands of goldenrod expanded. Stretches of that white-flowered aromatic herb, the narrow-leaved mountain mint, swarmed with nectar-hunters—bees, wasps, flies, and butterflies. I never tired of watching the little pearl crescents wind along the aisles among the slender stems. Once at one of these stands of mountain mint, on a day spiced with dangerous living, I delicately stroked with a forefinger the backs of eighteen bumblebees while they were lost in the delights of nectar-drinking.

Anyone who wanders over some old abandoned pasture with a pocket magnifying glass, examining the flowers of grasses and the forms and colors of mosses and lichens and leaves, enters a whole new world of beauty. And to lie on your back in the grass on a summer day, looking up at the drifting clouds, is to become aware of all the tiny overlooked insect sounds rising from among the slender leaves and roots beneath you. There is more going on in every pasture, every swamp, every ditch than even the most attentive owner is aware of.

On days when I roam over our pastures, amid the fluid grace of grasses in the wind, I sometimes come upon long straight grooves running through the dense vegetation. They are part of the trails of Trail Wood—trails that are laid out and kept open by other feet than ours. These particular paths, crisscrossing in our meadows, are the runways of the woodchucks. We did not come to Trail Wood to kill things. Even the woodchuck, against whom almost every man's hand is turned, finds sanctuary here. Watching its activity on summer days, I often recall Ernest Thompson Seton's characterization in his *Lives of Game Animals*: "I never see one in my field without getting a thrill of pleasure, a surge of admiration enhanced by respect. I see in him a character simple and brave." Surrounded by a world filled with danger, it endures. It thrives in spite of its enemies. It is doubtful, according to Seton, if there is any other animal of its size as numerous in America today as the woodchuck.

John Burroughs, in his implacable lifelong war on these pasture marmots, killed as many as eighty in one season in the area around his ancestral farm in the western Catskills. Yet half a century after his death, I drove along the road past those same fields and saw more than one of these universally persecuted animals sitting up in the sunshine at the entrance of their burrows. Woodchucks and gardens, I know, do not go together. Farmers, understandably, have a different viewpoint from mine. But it is our privilege to choose woodchucks. At Trail Wood they are welcome to all the grass they eat. They amply repay us with hours of interest spent in watching their activity.

We see them cleaning out their burrows in the early spring. The flickers come back and the woodchucks do their housecleaning. We see them stretching like kittens in the May sunshine—putting their forepaws far forward and extending to the limit each hind leg in turn. We see young woodchucks climbing up the slanting trunk of a rough-barked wild cherry tree, surveying for a first time the wide green world around them. We see them in late summer, floundering like seals as they rush toward the protection of their burrows. We see them carrying underground mouthfuls of dry leaves and grass for their winter nests. We see them sitting up in the autumn mist like smaller brown bears—the meadow bears of our open fields. We see them in the late fall, almost square, layered with fat, bulging with the fuel of hibernation, holding between their forepaws the red globes of fallen apples.

We remember special animals. There was one newly emerged baby woodchuck that ended one of its first days above ground with its stubby little tail a round mass of burdock burs. Then there was the half-grown woodchuck that tightroped across the top pole of a barway, repeating the performance several times as though exhibiting a parlor trick.

In the scale of animal intelligence the marmot is rated low. But in everything concerning its welfare it has all the sagacity needed for its survival. Its paths demonstrate that, without benefit of Euclid, it understands that a straight line is the shortest distance between two points. Around its burrow it usually leaves the grass and herbaceous plants high, screening the entrance, while it—like the fox that raids the poultry yards away from where it lives—feeds farther afield. Whenever possible the portal of its tunnel is framed in with roots or rocks that thwart predators trying to dig it out.

In things vital to its welfare it learns quickly. Once I called to a woodchuck feeding some distance away beyond a stone wall: "Hey there, Woodchuck!" It recognized me. I was familiar. It was far enough away to feel safe. It paid no attention to me. Then I barked like a dog: "Arf! Arf! Arf!" It took no chances and scuttled for its burrow. But the following day when I barked again it merely lifted its head and looked around, then went on with its feeding. It had been fooled only once.

In protecting its ever-endangered life, a woodchuck relies mainly on its eyesight. Because its eyes are set high on its head, it is able to survey its surroundings from the mouth of its tunnel without exposing much of its body. Out in the open it never feeds for long without stopping, sitting up, and carefully scrutinizing all the area around it. The closer the source of possible danger the oftener it repeats this performance. Several times I have checked on the interval between such surveillances. Always I have had to have recourse to the second hand of my watch. One woodchuck that saw me standing quietly about eighty yards away kept track of any change in my position by sitting up and staring intently in my direction at intervals of only seven or eight seconds. Another, feeding several hundred yards away in the open pasture, looked around it at longer intervals, about four times a minute. Its margin of safety was greater and it appeared to know it. Rarely, when feeding, does one of these marmots go more than a hundred feet from the safety of its burrow.

It is the young animals, appearing from their burrows each spring, that are less wary, more playful, more curious, and more fun to watch. They never tire of wrestling in the sunshine. We see them, day after day, endlessly trying out new foods. We have watched them bending down the stems of hawkweed to nip off the orange flowers, consuming the yellow blooms of the field mustard, eating the yellow-and-lavender-spotted white catalpa blossoms scattered on the grass. By the carriage stone, each morning one summer, a small woodchuck used to reach up and pull down one of the reddish-purple "popcorn balls" formed by the flowers of the common milkweed. It seemed to favor these blooms as a kind of appetizer before the heavier meals of the day.

One spring morning Nellie noticed a woodchuck consuming a bright golden dandelion flower. It looked so good she tried one. It *was* good. I tried one, too, picking a new flower, yellow to the center and free of small insects. It had a sweetish taste at first, then a dash of bitter, and finally a rather pleasant long-chewing quality. With us, ever since that day, eating a dandelion flower has become one of the rites of spring.

I have never seen a woodchuck drink. No doubt plant juices, dew, and rain provide it with all the liquid it requires. I remember watching a woodchuck eating lilac leaves in the rain and, another time, two that had only recently appeared from the burrow in which they had been born sitting almost side by side in a warm late-spring downpour, one eating dandelion leaves, the other the leaves of wild lettuce. Sunflower leaves, sassafras leaves, nettle leaves, plantain leaves, milkweed leaves all are relished by woodchucks. Later in the summer they add to their diet fruit of many kinds. A friend of mine was amazed to see a woodchuck getting wild grapes from vines almost at the top of a red oak tree. Each year quantities of small green apples are devoured by our woodchucks, as near as we can tell with no stomachaches in consequence.

The general idea that unless woodchucks are controlled they will multiply and overrun a farm has not worked out so far as our experience is concerned. None of our woodchucks is killed, yet we have about the same number each year. They are territorial animals. Each groundhog guards its own area. So natural pressure spreads them out and prevents a concentration of population.

Among woodchucks there is no Share-the-Wealth Plan. One large male, for several years, lived under a wall beneath an apple tree. In late summer days, when the ground was littered with red fruit, a particularly small woodchuck that lived beside the lane would arrive stealthily in search of apples. But always it was sent racing back to its burrow by the rush of the guardian male. We grew concerned over its lack of fat. Would it develop enough reserves to carry it through the winter? To build it up, I remember, we carried several pails of apples and dumped them in the weeds beside its burrow. In the course of another autumn we saw a second undersized animal continue eating long after all the other woodchucks had disappeared. It seemed to sense its need for additional fat to keep the machinery of life running in low gear during its months of hibernation. Day after day it consumed cracked corn we had put out for the mourning doves. We watched it in its race against time and when we saw it no more, and during the months of snow that followed, we often wondered if it would last the winter out. But it did. Spring came and it was among the awakening woodchucks that appeared in the sunshine to find the grass of the meadows green once more.

Where, to the north, the Starfield swells to its highest point, we call the elevated ground Nighthawk Hill. This name commemorates a rare adventure of ours in the sunset of one of our earliest years at Trail Wood. Toward the end of an August day, we had emerged from the woods and were starting across the meadow when I glanced up. Overhead, like an apparition of beauty in the sky, a great wheel or funnel of birds—more than sixty, gray, slim-winged, with silvery-white patches that shone in the tinted rays of the sunset—was turning without a sound. We threw ourselves down on the ground, warm at the end of that sultry day, and, lying on our backs, looked up at the buoyant, lightly loaded birds, at this aerial whirlpool of wings revolving above us. We saw the wheel of streamlined bodies, of graceful wings, turning against the delicate background of high feathery cirrus clouds flushed with pink in the sunset light.

More than once we had seen broad-winged hawks on migration wheeling in this manner. But this was our first encounter with a "nighthawk circus." For fully a quarter of an hour the whirling birds turned in unison above our pasture fields. Then they drifted slowly away toward the south. Since that August day, we have seen many nighthawks veering over the Trail Wood meadows but never again that revolving wheel of life turning so close above us. Once and only once, only on Nighthawk Hill, have we been so fortunate.

I remember hearing of a New England farmer who had the luck to witness the spectacle of a nighthawk circus over his land. The one thing he recalled most vividly was that the whirling vortex had come so low he had been able to knock down some of the birds with a stick. How far back in history runs this vicious thread! As in a reflex action, like the cat that cannot resist pouncing on a string drawn before it, certain men in every age have responded by hitting, maiming, killing every wild creature that has come close.

This age-old reaction is changing gradually—all too gradually. At least no longer do correspondents write as one correspondent did to *Forest and Stream*, in 1885, to announce the discovery of why nighthawks had been created. A great light had dawned in the darkness of his mind. His revelation was that nighthawks had been put on earth because: "Their rapid and singular flight makes them a difficult target for young sportsmen to practice on."

Years ago, in *Mosses with a Hand-Lens* by A. J. Grout, Nellie came upon a description of *Pleuridium subulatum*, a fine furlike species bearing minute globes of green. "Down among the tufts of grass in dry and sandy fields in early spring," Grout writes, "one can find soft silky tufts of green containing innumerable little green spheres like emerald dewdrops. These green spheres are the capsules nestling among the leaves because of the shortness of the setae. The illustrations can give no idea of the beauty of a dense tuft several inches square, fresh from the field, wet with the spring snows and rains." How such a description, come upon among dry pages of technical botanical terms, opens the door and lets the spring come in!

Ever since she first read those words, Nellie had longed to come upon the subject of this description, to find the growing plant, to see this tiny moss—only an eighth of an inch or so high—with its living dewdrops of green. And on an April day in our west meadow she did. As we examined it our pleasure in its elfin beauty was all the greater for knowing that, on some unknown day long before, this same loveliness in miniature had stirred a scientist to the poetry of feeling. Another April came and Nellie looked in vain for the emerald dewdrops. Never again could we find them. Like the nighthawk circus above Nighthawk Hill, they represented an adventure in our meadows that was deeply enjoyed, long remembered, but never experienced again.

In that same portion of the western pasture, hardly have the snows melted in the spring each year when one of the most profoundly moving events in the life of our meadows commences anew. All through the latter days of winter we look forward to it with mounting anticipation. Then at the beginning of dusk, usually at the close of some March day, we hear once more the magical flightsong of the woodcock.

Its performance begins with the brown chunky long-billed bird walking about in the open field, turning this way and that, uttering again and again a buzzing nasal "Peent!" Often it is hidden in the dusk but on the occasions when I have been able to observe it I have noted how it lifts its wings, hunches its shoulders and jerks its head at every "Peent!" The call seems now far away, now close at hand, according to the direction the bird is pointing. Then there is a moment of silence. It is followed by the winnowing sound of its wings and we see its dark little form speeding in a wide climbing curve against the light of the sky. Often it passes directly over us as we stand on the back-door steps.

Higher and higher in great sweeping circles it mounts above the pasture. We follow with our eyes its retreating form, often losing it in the sky. At the height of its ascent the song begins. The sweet frail twittering sound at times seems to come from all directions, the notes to shower down around us. And while the song goes on it is joined by a quavering musical strain produced by three stiff narrow feathers at each wingtip. They vibrate in the wind as the bird plunges, veering wildly, falling through the sky like a gust-blown leaf. The end comes abruptly—an almost vertical descent to the darkened meadow. Then the "peenting" calls begin once more.

There is always something particularly moving about the ecstasy of this dull-appearing brown and dumpy bird, this feeder on earthworms, this boggy-ground dweller, as it mounts up and up into the sky and then plunges with its wild gyrations back to earth again. It seems a spirit unfettered for a time, transcending its ordinary days, attaining a superlative moment. There is something symbolic about its flight. We too feel a lift in spirits as we follow it with our eyes as it speeds across the sky. How fortunate we are to have woodcocks to sing in the twilight outside our kitchen door.

The Time of the Woodcock—that wonderful time of spring's beginning! Each year we experience the same thrill. Once more the woodcock! Once more the sweet twittering song coming down from the darkening sky! Once more the falling-leaf descent and the final plunge to earth! I find I wrote in my journal on April 3, 1969: "Woodcock up again in the dusk. I wish I could enter this note every day in the year!"

Some seasons the woodcock sings before the first chorus of the spring peepers. It is the early voice of this returning migrant that marks the winter's end. During successive years we have seen our woodcock ascending through falling rain, through ground mist on the meadows, from fields powdered with late snow, under crystalline skies with the full moon shining and under skies so dark and stormy we could never catch sight of the singer in the air. At such times, when the fires of the breeding season are burning bright and the male is seeking to attract a mate, weather has little effect.

One tempestuous day at the beginning of April is still vivid in my mind. With a spitting of icy rain and a gale of wind, a cold front struck in the afternoon. The thermometer dropped ten degrees in an hour. Toward evening a flurry of snow was followed by clear brilliant skies. About seven o'clock, bundled in heavy clothing, I sought protection where the west wall joins the garage overlooking the pasture. The mercury stood at thirty-two degrees F. and gusts were reaching fifty miles an hour. Only a little past full, the moon spread its pale silvery light across the open fields and over the dark lashing tops of the trees. While I crouched there, over and over the woodcock climbed into the cold windswept sky, riding the great gusts, to come tumbling down again. Two or three times during these storm-flights, in momentary lulls in the pounding surf-roar of the gusts, amid all this tumult, my ears caught small tender fragments of its aerial song.

During the weeks of early spring our days at Trail Wood are bounded—morning and night, dawn and dusk—by the song of the woodcock. Always crepuscular, they occur twice each day, when the earliest light is breaking as day is beginning and when the earliest darkness is falling as the day is ending. Their purpose is twofold. They attract any unattached female flying by the polygamous male. And they warn other males that a territory is occupied. At times, on calm evenings, we will hear as many as three males singing above different pastures around us.

The fourth year we were at Trail Wood, on a late-March evening, we were listening to two males, one in the west pasture and one in Firefly Meadow. First one would go up and then the other. Suddenly, while the West Pasture woodcock was "peenting" on the ground, the bird from Firefly Meadow passed us flying at top speed and following a beeline for the calling bird. As it went by we heard a new sound, a kind of low, staccato "a-a-a-a-a" or "cac-cac-cac-cac-cac" like a child imitating a machine gun. A moment later we saw the two birds, flying low and fast, one in close pursuit of the other, go streaking away for the North Woods. Several times since we have heard the same sound. It is evidently a note of threat or intimidation, perhaps produced by snapping the mandibles together. Each time the bird was rushing toward a rival.

With us the song of the woodcock commences anytime from mid-March to the early days of April. It varies from year to year. Also variable is the number of days the flights continue. In 1964 a male ascended from the West Pasture in the dusk of thirty-five successive evenings. Five years went by before that record was eclipsed. Then, in the spring of 1969, the mark was raised to thirty-nine nights, extending from April 1 to May 9. Two years later, in 1971, the present all-time record for Trail Wood was established. The spring that year was particularly backward. In its outdoor effects it was calculated by the weather bureau to be a month behind schedule. But the woodcock's first flight from the West Pasture occurred on March 18. Night after night—in rain, in mist, in wind, in moonlight—the male sang on. April passed and May arrived. Its serenades continued. Was its song, like the poet's verses of unrequited love, the result of failure to attract a mate? In many species of birds, the males outnumber the females. The month of May was almost over, the twenty-sixth day had arrived, when the final flight occurred. We stood in the warm lilac-scented dusk while above us, for a last time that year, we heard the liquid notes of its aerial song. Then, invisible in the gathering darkness, the singer landed. For a long time we heard it "peenting" on the ground. It took off again. But this time there was no mounting into the sky. We heard the sound of its wings as it flew away, low above the pasture, straight for the dark North Woods. The sixty-nine nights of its aerial singing had come to an end.

The close of these flights was the ending of an era in the history of that year. At Trail Wood we use clocks and calendars. But the important divisions are marked in our minds in a different way: the Time of the Katydids, the Time of the Melting Ice, the Time of the Fireflies, the Time of the Turning Leaves, the Time of the Woodcock's Song. The calendar of nature has other divisions than weeks and months. It notes a long succession of events annually recurring.

A day or so after the dusk songs of the woodcock ended we stood on the kitchen steps, listening once more in the deepening twilight. We heard the clear, pure, simple strains of the field sparrow. We heard the robins, the towhees, the catbirds, the veeries, and the wood thrushes of the darkening woods. But the wild meadows seemed lonely. Something was lacking. No more the nasal "Peent," no more the winnowing wings of the woodcock's takeoff; no more the ecstasy of its falling-leaf descent to earth. In all the consecutive dusks from mid-March almost to the end of May it had provided us with a thrilling climax to the day. We had heard its final call. We had been present at the precise moment when the fires of spring suddenly were extinguished. We had caught, on that final evening, the sound of its wings growing fainter as it flew to the woods and its natural feeding. Taking off, going faster, gone—this was a major turning point for this strange crepuscular bird, the end of all those serenades at last. We came in that evening and closed the door, wondering where, in all the darkness that had enveloped our woods and fields, the singer, now silent, pursued its ordinary way.

CHAPTER SIX

The Village on the Hill

One of Chekhov's short stories is set in a village in a ravine. Whenever those who passed by asked what village it was, they were told: "That's the one where the deacon ate up all the caviar at the funeral." This single unimportant event was the only thing ever remembered about the hamlet.

Our village, whose white church spire we can see rising above the trees two miles to the south, is doubly the reverse of Chekhov's. Instead of nestling in a ravine, Hampton occupies a high point in the landscape. Instead of being a community where nothing has happened, its history is replete with incidents of historical and human interest.

It was the home of the famous "Twelve Sons" of the Foster family who fought with their father in the Revolution. Their united service is said to have exceeded that of any other family in the American Colonies. The village also was the home of Abijah Fuller, who drew up the lines of fortification on Breed's Hill at the Battle of Bunker Hill. He was one of seventeen cousins from a single Hampton school district who enlisted in the Continental Army. During the French and Indian War that preceded the Revolution, Robert Durkee of Hampton rose to the rank of captain. Before the Civil War the village formed a link in the Underground Railroad that assisted Southern slaves in reaching sanctuary in Canada. There is even a legend of buried treasure here that has led to sporadic digging in the past.

For one day, in the 1840s, this small rural community overlooking Little River was designated the capital of Connecticut. At the time its most celebrated inhabitant was the governor of the state, Chauncey F. Cleveland, brother of the man for whom Cleveland, Ohio, was named. According to the story as I have heard it, Cleveland was confined to his home in the village by an injury at a time when his signature was urgently needed on a document that precedent required must be signed in the capital. This difficulty was surmounted by the simple expedient of declaring that Hampton was the capital for that one day.

In the north end of the town lies "the House the Women Built" and in the south end the historic Curtis Tavern, which John and Dorothy Holt transformed into a private dwelling. This hostelry was celebrated for its "spring" dance floor and, in earlier times, for the name printed on its signboard. The original proprietor, when he came to paint the sign, asked his wife for an appropriate name for the tavern. She pointed out that his money was sunk in the venture and he might well lose it all, and the most appropriate name she could think of was "Man Coming Out the Little End of the Horn." So he painted that legend and beneath it a picture of a large horn with a small figure creeping from its smaller end. By this unusual name the tavern was widely known in stagecoach days.

Twenty-three of the fifty states of the Union contain communities called Hampton. Our Hampton was among the first to receive the name. Its past stretches back through more than two and a half centuries to the first settlers to arrive from Massachusetts. As early as 1709, David Canada migrated from Salem and established himself on land beside Little River. He was the first to build a house and the first to open a tavern in the area. It was his name that was given to the parish—Canada Parish—when a charter was granted in 1717. Five years before that, land on Hampton Hill—then known by the name of Appaquage Hill and later, for a time, as Chelsea Hill—was opened to purchasers. Its attractions were listed as: "Soil good. Land cheap. Situation pleasant. Outlook commanding." Just twenty years before our house at Trail Wood was built the Connecticut Assembly, on October 2, 1786, passed an act establishing the Town of Hampton as an administrative unit within the County of Windham. Its population then was 827.

During the final years of the eighteenth century, that population grew, largely through activity along the banks of Little River. This pleasant country stream, famous for trout, provided the water power for an increasing number of small factories. They produced buttons, safety pins, spectacles, rakes, plows, hats, German silver spoons. At one period the stream supplied power for five grist mills, seven sawmills, and three shingle mills as well as for cider mills, fulling mills, a potash mill, and a cotton mill with 1,000 spindles. When the earliest machinery for grinding grain was set in operation, Indians came from miles around to sit all day beside the river, fascinated by the turning mill wheels. A tradition of craft work continues in the village into the present. Various homes in Hampton produce ceramics, weaving, handcrafted jewelry, candles of original design, and wooden toys.

In 1790 the census of that year showed that the number of inhabitants of the town had risen to 1,333. Among them was one slave, the property of the Reverend Samuel Moseley, whose tenure as minister in the village extended over a period of more than fifty years. But the heyday of the little factory passed. Conditions changed. The population decreased and Hampton became what it still is, a rural village that, on all sides, quickly merges with the green pastures and wooded hills of the surrounding country.

The eminence on which it stands is part of a line of hills running north and south and roughly paralleling the course of Little River. Covering its summit like a coat of armor is a layer of harder rock overlying the softer material the river has eroded away in forming its valley. In New England, in Indian times, hilltops were favored places for villages and farms. They provided strong positions for defense and vantage points overlooking the country around. When three pioneer families—the Kimballs, the Grows, and the Fullers—moved together from Ipswich, Massachusetts, to Hampton, they settled on the three highest elevations in the area. One is still called Kimball Hill, another Grow Hill. The third, which long bore the name of Fuller Hill, is now known as Sunset Hill. Its 812-foot summit is the highest point in Hampton.

If you look down the road that descends steeply past Marjorie Medary's house into the valley at the north end of the village, you see lifting beside the river, 350 feet below, a smooth rounded knoll, a kame deposited by the glaciers. From top to bottom it is covered with white gravestones. This, Hampton's North Cemetery, has been used as a burying ground since the days of the earliest settlers; perhaps by Indians before them. In former times, before glacial action was understood, the origin of this huge, isolated mound in the valley caused endless speculation. One of the wilder suppositions was that the mound had been produced by pioneers hauling sand and gravel to the spot in wagons pulled by oxen.

In its gravestones this village burying ground records the martial history of the United States. Here are buried veterans of the French and Indian War, the Revolutionary War, the Civil War, the Spanish-American War, the First World War, the Second World War. Among its graves is that of the father of the "Twelve Sons" of the Revolution, Peter Foster.

On July 4, 1826, in celebration of the fiftieth anniversary of the signing of the Declaration of Independence, forty-two white-haired veterans of the Continental Army, all from the village of Hampton, marched the length of the main street. Many were dressed in their old uniforms. They were led by Abijah Fuller and among the fifers who played martial airs of '76 on that day was Joseph Foster, one of the "Twelve Sons." A number of these Hampton veterans lived to be nearly 100. In noting this fact in her *History of Windham County, Connecticut*, Ellen D. Larned adds: "The pure air, generous living and social amenities of this pleasant town were eminently favorable to health and longevity." Apparently they still are, judging by the number of those who have reached their eighties and even their nineties while we have been here.

The wide street between the white houses and shaded lawns, along which these veterans marched that July day in 1826, was originally known as Great Street. A green once extended down the center of its length. It is now hard-surfaced and part of State Route 97. However, in outward appearance the hilltop village running along it has changed surprisingly little. Even today, if you follow the course of that parade, you see some of the same buildings the veterans went by almost a century and a half ago. At these old houses, in earlier generations, traveling shoemakers used to board for several days at a time while they completed shoes for the entire family—a process known as "whipping the cat."

Walking as the parade advanced, from north to south along the street, you glimpse on your left a small charming cottage set between larger dwellings. It is called "the Nutshell." Beside it is a structure that has been altered and added to at various times in its long history. Until the 1940s it housed the Chelsea Inn. For more than 200 years hospitality was provided at this site for wayfarers passing through Hampton.

Diagonally across the street, the second oldest Congregational Church in Connecticut lifts its high white spire, a landmark for miles around. This church was designed and built in 1754 by Thomas Stedman, a native of Hampton who became famous throughout eighteenth-century New England as an architect of meeting houses. At the time he was only twenty years old. Later, when he was nearing sixty, he supervised repairs on this building he had designed in his youth. The steeple was added in 1792; 146 years later, during the hurricane of 1938, the steeple and the half-ton bell it contained crashed to the ground and had to be replaced.

In the record of the innumerable events that took place beneath that steeple there is noted a public apology made by one parishioner after he had declared: "I had rather hear my dog bark than Mr. Billings preach." This far northeastern corner of the state has had its share of the sharp-tongued and stiff-necked—including the litigious woman who, when told by an exasperated judge: "There is enough brass in your face, madam, to make a five-pailful kettle," shot back: "And sap enough in your head, your honor, to fill it!" and the Hampton boy who complied with the teacher's request for a written apology with: "If I have done anything that I am sorry for, I am willing to be forgiven."

Just to the south and next door to the church stands the Fletcher Memorial Library, established in 1924. To its west, on Cedar Swamp Road, the Catholic church is nearing its hundredth year. The library occupies a large, formerly private home. As early as 1807 there was a library in Hampton. According to old records it contained "over 100 volumes." It was succeeded by other libraries. One, after being in existence for three years, was discontinued and the books were sold. Later it was reorganized and the books bought back again. Nearly 15,000 volumes fill the rooms of the present library. Meetings of the board are relaxed and cover a wide range of serious and humorous topics. I remember at the first board meeting I attended one item of business was a check that had come in to the librarian, Eunice Fuller. It represented payment for a borrowed book that a puppy had chewed.

The front windows of the library look across the street to the village store, now vacant. It was here that the first telephone in the community had been installed. When we came to Hampton, the social and friendly proprietor was Francis Wade. I never met a man who appeared to enjoy being a storekeeper as much as he did. "If only," he once said to me, "I could break even!" Caught between unpaid bills and the competition of supermarkets, he continued fighting his losing battle for several years thereafter.

Until 1949 U.S. 6 climbed in a steep ascent up the escarpment from the river valley and turned south along the main street. Then the present bypass, eliminating the long climb, diverted traffic farther to the south, carrying the flow of trucks and buses and cars, the noise and fumes, away from the village. Where the old route reaches the top of its ascent just south of the store, it curves around the one-time home of Chauncey F. Cleveland. In spite of years of neglect, it is still architecturally one of the most beautiful houses in Hampton. During his years as governor, Cleveland furthered progressive and humane legislation. He worked for the passage of laws to regulate child labor, eliminate imprisonment for debt, and make provision for the care of the insane poor. His father, Silas Cleveland, who as a boy had been captured by the Indians and carried to Canada, spent his last days in Hampton.

Over the way, on your right as you continue south, you observe a large white house with double porches, one for the first, one for the second story. This, for many years, formed the home of two remarkable sisters, Helen Mathews and Annie Edmond, the blind naturalist of whom I will have more to say in a later chapter. Both had been born on a farm close to the origin of Little River. Two of their earlier ancestors were John and Priscilla Alden and they were related to the bride responsible for "the House the Women Built." With keen minds and unusual memories, they spanned in their personal recollections more than eighty years of Hampton history. A number of families in the community have been associated with the town for 100 years or more. The Jewetts and Pearls and Fullers go back for more than 200 years.

Progressing down the line of houses on this side of the street, you pass a dwelling built about 1755 by John Brewster, a descendant of Elder Brewster of the *Mayflower*. This was the home, at one time, of the celebrated illustrator of magazines and books, Florence Nosworthy. Next door, in the 1800s, lived Dr. Dyer Hughes, who practiced medicine in Hampton for more than fifty years. He made the rounds of his early patients on horseback and, when he first arrived, his fee for a house call was twelve and a half cents.

That center of village life, the post office, now presided over by Charles Fox, occupies a shelf of land overlooking the valley on the other side of the street. It is here you meet your neighbors, hear of the return of travelers, learn who has seen the first robin or heard the first geese going north in spring. Not far beyond, old Route 6 used to turn to the right and leave the main street behind. It was near this spot that the forty-two veterans of the Revolution, in 1826, ended their parade.

If you continue on from here, you see on your left the hall of Little River Grange, organized in 1885. Beyond it the grade school of the community occupies open land sloping toward the east. The vast sweep of valley and hills that extends away below forms one of the finest views in the region.

This, then, is the village on the hill whose church spire we see above the trees. We remember it in the October sunshine when sugar maples all along the street are clothed in their autumn colors. We remember it in the winter dusks when lights from the windows are reflected on the snow. All around its edges there is wilder land. One winter a pheasant hen took shelter under the back porch of the seventeenth-century house where Harold and Grace Stockburger lived. Wood ducks have nested in hollow apple trees in back yards overlooking the valley. At one time a farmer, working in fields near the river, used to use a chimney on the roof of one of the village houses as a sundial, deciding by the position of its shadow when it was time to go in for lunch.

Still retained here are a number of pleasant old-time village ways. Each year the Little River Grange puts out a calendar that lists birthdays as an aid to sending greetings. At a meeting of the Hampton Antiquarian and Historical Society, one winter night, cream was whipped up in cider to produce "sylabub," a rich, frothy, sweet-flavored drink dating back to Colonial times. Town meetings are still held. One long-standing custom of the village, believed to be two centuries old, is still observed. Quietly, without publicity or general solicitation, a collection is taken up and presented to the next of kin when a death occurs. The recipient may donate the money to some charity as a memorial or he may use it in whatever way he desires.

During our first year in Hampton I left my hat behind on a visit to the south end of town. The next morning I found it in my mailbox. The rural mail carrier, on his rounds, had brought it home. This friendly, helpful attitude of the village is possible only in such less overcrowded, less regulated, less high-pressured parts of the land. In a city simple acts of kindness seem more difficult. Individuals there are more tense, suspicious, on guard than in the village. A friend of mine tried to help an old lady on a New York City bus. She hit him over the head with her umbrella.

In general, an air of unstrained calm, of live and let live, is a feature of our village life. Hampton is the kind of place that provides such quiet notes amid the generally raucous din of the newspapers as: "Town Clerk, Mrs. Margaret Fox, reminds bee owners their hives must be registered in her office by October 1."

Like every other community on earth, our village has its problems, its quarrels, its crimes, its high taxes, its alarms and excitements, its hardships and tragedies. But those who have been born in Hampton and those who have long resided here seem to have developed a special affection for it. In her *Folklore and Firesides of Pomfret, Hampton and Vicinity*, Susan Jewett Griggs expressed their viewpoint in these words: "Hampton is one of the most beautiful towns in the state. God did something for Hampton, in rolling hills, valleys and streams, that makes us feel—this is our home—it's Hampton."

CHAPTER SEVEN

Birds of an Old Farm

During the long days of spring at Trail Wood—as in "My Old Kentucky Home"—"the birds make music all the day." Before five o'clock in the morning their dawn chorus begins. We awaken to the voices of wood thrush and veery, bluebird and robin, purple finch and catbird and scarlet tanager, yellowthroat and field sparrow, rose-breasted grosbeak and northern oriole. From the weed tangles along the stone walls of the pastures comes the clear, carrying whistle of the bobwhite; from the brookside trees the "switch-you" song of the chestnut-sided warbler; from across the pond the hoarse repeated call of the great crested flycatcher. Between forty-five and fifty species of wild birds, almost every spring, nest within singing distance of the house. Twice barn swallows have constructed their mud nests and reared their broods on top of the projecting floodlight just outside the kitchen door.

At the end of his voyage to the New World, Christopher Columbus reported to King Ferdinand and Queen Isabella that, among the islands of the Caribbean, he had found himself surrounded by "birds of a thousand sorts." During these days of full birdsong in the spring, there are times, on this New England farm, when we experience a similar impression.

When we arrived at Trail Wood, one of our earliest ambitions was to record as many species of birds for our farm as Gilbert White listed for his English parish of Selborne. In 1774, in a letter to Thomas Pennant, White reported the total for his Hampshire countryside was "more than 120 species." We began our list the day we arrived.

The first bird was a phoebe, nesting in the garage, the second a red-shouldered hawk, the third a flicker, the fourth a towhee, the fifth a meadowlark. These early birds came fast: mourning dove, chimney swift, Cooper's hawk, barn swallow, catbird, woodcock, nighthawk. The ninetieth bird on the list was a pine siskin attracted by food scattered on the snow in January. The hundredth passed high above us on the eleventh day of May in 1961. Shaped like a cross, moving swiftly under a cloudy sky, this late migrant was a loon returning to some remote pond among the forests and mountains of the north. Only on this one occasion have we caught sight of this most ancient of North American birds winging its way across our sky at Trail Wood.

Beyond the hundred mark additions came more slowly. Number 101 was a Canada warbler, 102 a magnolia warbler, 105 a spotted sandpiper, 110 the first mallard duck to alight on our pond. A hermit thrush in the woods, a bank swallow flying overhead, an American bittern by the pond raised the total toward 120. It was on the sixteenth of June, in our second year at Trail Wood, that we added our hundred and twentieth bird, a Nashville warbler singing in an elm tree below Juniper Hill.

Since then the count has crept upward to equal and surpass White's Selborne list. On October 28, 1964, when Edward and Camille Dodd were visiting us, Nellie and Camille caught sight of bird number 128, a pileated woodpecker, near Whippoorwill Spring. Our list now stands at 144. One of the last additions was a coot that arrived in the fall, fed with the mallards during more than a week of fine weather, and took off in the night ahead of an all-day storm. Another was a rare Harris's sparrow that had gone astray during its migration south. Still another was the most surprising of all, the last bird we expected to see at the time we saw it. It was large, white, yellow-billed, a great egret stalking about beside our pond during a brief thaw in mid-January.

Among these many birds of many species certain individuals—seen during odd or special moments in their lives—stand out in memory. One was a brown thrasher on the wall under the hickory trees. When I first caught sight of it, it was stuffing food into the gaping mouth of a fledgling cowbird that had just left the nest where its egg had been deposited. I saw the thrasher arrive with a large angleworm. It pushed it into the young bird's mouth. Most of it hung down on either side of its bill. Unable to swallow it, it dropped it on the wall. The brown thrasher snatched it up and thrust it into the open mouth once more. The results were the same. Over and over, five, ten, fifteen times, the process was repeated. After the fifteenth time the brown thrasher swallowed the worm itself.

A robin with a woolly bear provides another memory. The bird picked up the spine-clad caterpillar in the grass. For a full five minutes, before it swallowed it, we watched it pound the larva on the ground, turning it this way and that, breaking off the stiff hairs that covered its body. Then there was the forlorn little myrtle warbler that Nellie discovered on a burdock plant one autumn afternoon. Somehow it had become entangled among the burs. How long it had been imprisoned there, with one wing held outstretched, we have no way of telling. But in its fright, as Nellie reached out a hand to help it, it redoubled its struggles and tore itself free.

I also remember how, one April morning when bluebirds were nesting in one of the weathered bird boxes near the brook, I saw a flicker, in the full flush of his springtime exuberance, swoop down, cling to the side of the nesting box and commence a trip-hammer tattoo with its bill on the wood. The hollow interior, like a hollow limb, reverberated with the sound. The loud rolling signal of the woodpecker carried across the yard. But it lasted for only a few seconds. Out rushed the indignant bluebird and drove the flicker away.

Day after day and week after week, one year, a male song sparrow fought its reflection in the rear windows of our middle shed. Beyond the lilac bush three sheds are linked together in a line. The first holds our car, the second our fireplace wood in winter, and the third, a former cowshed, forms a combined workshop and storage place. The windows at the rear face west onto the overgrown edge of the pasture. Apparently, in this particular year, a pair of song sparrows raised two broods in quick succession close to the buildings. Guarding his nesting territory against intruders, the male spotted his reflection in the windows of the middle shed, mistook it for an invading male, and flew to the attack. At times this bewitched bird shifted from one window to another, fighting as many as four imaginary enemies at one time.

I first noticed it on May 18. It was whacking the glass with its bill with such force that I heard the sound 100 feet away. I looked at the second hand of my wristwatch and counted the strokes. At their slowest they were coming at the rate of more than one a second; at their fastest the bird was raining ninety strokes a minute on its image on the glass. I wondered how long it would be before its bill wore away. I was still wondering that weeks later. For through the rest of May and all through the month of June and on through much of July the little bird continued to attack the panes of glass. As early as seven-thirty in the morning and as late as seven o'clock in the evening I would find it belaboring this phantom rival, its unvanquished enemy, its own reflection.

Never before nor since have I encountered so pugnacious a sparrow. Throughout the day it returned at intervals to the attack. On the twenty-seventh of July I saw it make its last onslaught. On that evening I heard it singing on the ridgepole of the shed. It seemed master of all it surveyed. For seventy-one days it had been engaged in its frantic one-sided assault. How many times, during those days, had it struck its bill against the glass? Its shadowy enemy was illusory but the hard surface of the pane was not. When I calculated the average number of strokes at fifty a minute and the total time spent at the windows in the course of a day at two hours—both of which figures seem extremely low—I discovered that during the ten weeks of its long enchantment, the song sparrow had battered its bill against the windowpanes in excess of 400,000 times.

All through the spring and summer we leave one of the wide doors of the middle shed blocked open for the phoebes to come and go. Once when a brown thrasher flew inside the shed and blundered about from window to window, unable to find its way out again, I caught it in a butterfly net and released it outside. Barn swallows and robins, as well as phoebes, have nested in the middle shed. One June, when a robin nest was overflowing with young almost ready to fly, one of the nestlings prematurely launched itself out and fluttered down to a corner of the shed. I learned again, that morning, the mistake of trying to put a baby bird back in the nest in such circumstances. When I attempted it, all the rest of the brood exploded in different directions. No amount of putting them back in the nest would make the nestlings stay there. A link had been snapped. The urge to stay had been superseded by the urge to leave the nest. In trying to return the first bird, I had only increased the labor of the parents. Instead of feeding the brood together on the nest, they now had to feed the young birds separately on the floor.

Although we have encountered the dense, feathery cushions of Schreber's moss only in one place at Trail Wood, it is this moss that appears every year in the nests of our phoebes. This same primitive plant, I am told, is used in the nests of these birds along the coast of Maine. One spring a pair of phoebes building a nest in the middle shed reduced the labor involved by taking a short cut. Instead of gathering fresh material from the woods, they stripped the dry last-year's Schreber's moss from a nest in the storage shed next door. Curiously, one of the most unusual nests we have encountered in Hampton also has its link with this species of moss. A mile or so from us, one year, chickadees nested in an unused rural mailbox, coming and going through a narrow opening in the front. When the box was opened it was found to have the floor covered deeply with a soft wall-to-wall carpeting of green. This carpet was formed entirely of Schreber's moss.

Not only are many species of birds expert botanists at nesting time, but they all—even birds that have never made a nest before—exhibit an inborn capacity for selecting the most suitable material from all the plant fibers, twigs, hairs, and grasses around them. We see northern orioles ripping long tough strips from the dry stems of last year's milkweed plants and chipping sparrows hunting for hairs shed by animals. Sometimes the nests of these sparrows are made largely of fragrant sweet vernal grass. Occasionally one will be lined with the reddish hair of a deer, at other times with long hairs from a cow's tail found caught on barbed-wire fences. I have one unusual nest given me by Bert Inman, who lives at the south end of Hampton. It is lined with the soft blond hair of his young daughter, Jane. When, on a warm spring day, her mother cut her hair under an apple tree in the yard, a chipping sparrow collected the discarded material as a windfall in completing the small cup of its nearby nest.

Along Veery Lane, on an April day, Nellie came upon a chickadee hopping this way and that over the ground. It was furtively gathering, as lining for its nest, small tufts of rabbit fur that lay scattered on the path. New materials, the innovations of technology, are increasingly finding their way into the nests of birds. Wood thrushes have made use of soda straws in place of twigs and great crested flycatchers have employed strips of cellophane in place of snakeskins. Several times sheets of plastic material Nellie tied to bushes in the woods to mark special wildflowers have disappeared. We suspect the flycatchers. Not long ago, when I opened one of our nesting boxes, I discovered several narrow ribbons of aluminum foil among the dry grasses of a tree swallow's nest.

Perhaps they were used as a substitute for feathers. For, whenever possible, such swallows line the inner cups of their nests with white feathers, most often those of chickens. Along the coast gull feathers are employed. Nellie and I, each spring, watch the birds toying with feathers in the air, dropping them while on the wing, swooping and catching them again in their bills. They never seem to tire of this sport as they work toward their nesting sites.

I am reminded of an amusing incident that occurred at the entrance of one of our bird boxes on a morning in May. Carrying a larger feather than usual, more than five inches long, a tree swallow flew toward the box. It held the feather crosswise in its bill. At the entrance hole, an inch and a half in diameter, it was brought up short. It circled, dropped the feather, caught it in a new position and tried again. Time after time we saw it approach the hole but always it was blocked by the feather held crosswise. It never did overcome the difficulty and finally abandoned its feather. However the tiny brain of the house wren has found the solution to the problem. We have watched these midgets come speeding in, clutching surprisingly long twigs in their slender bills. Without pausing, but with a quick sidewise jerk of their heads, they turn the twigs lengthwise at the last instant and pass through the opening with ease.

Each morning during the nesting season we crush up the shells of our breakfast eggs and scatter the fragments on the driveway outside the kitchen door. Within minutes birds alight to feed on these particles rich in calcium carbonate. Usually the first to arrive are barn and tree swallows. I remember one barn swallow that came back nine times —no sooner taking wing than circling back to feed again—before its craving for eggshells was satisfied. Aside from the swallows, a surprising variety of other birds are drawn to the fragments. They include kingbirds, bluebirds, northern orioles, house sparrows, great crested flycatchers, bluejays, flickers, and goldfinches. Whenever we have been able to tell the sexes apart, the visitors have always been females. Usually the last in the season to arrive are the goldfinches. It is normally well along in August before these late-nesting little finches begin to show an interest in our source of calcium carbonate.

Secretiveness is part of the nesting season for many of our birds. We notice how silent and furtive the bluejays become, how our chickadees and tufted titmice disappear entirely from around the house while they are raising their broods. Once, when a crow passed us carrying a stick to add to its nest, it saw that we were looking up. It realized it was observed. Immediately it dropped the stick and veered away in another direction. Similarly toward evening during the first week of July, a red-eyed vireo, near my writing cabin among the juniper and aspen on the other side of the pond, sought, in a transparent ruse, to deceive me and conceal the fact it was carrying food to nestlings. When I caught sight of it—and it caught sight of me—it was carrying a small white butterfly to its nest. It abruptly changed its course, alighted on a twig, peered at me intently for some time. When I remained standing, watching it, it swallowed the butterfly and then flew away, apparently convinced I had been misled.

This secretiveness extends to the careful elimination of all clues that might call attention to the presence of the nest. Eggshells left by the hatching of nestling songbirds are usually removed and dropped some distance away. The fecal sacs of the young are carried from the vicinity of the nest. We see tree swallows dropping them while flying over the pond and song sparrows transporting them to trees where they leave them attached to twigs. I recall one pair of chipping sparrows that nested on the lowest branch of an apple tree. The parent birds carried away the sacs in quick furtive flight. They always left them in the same place, deposited on the top strand of a barbed-wire fence.

I believe the most secretive bird I have ever known at Trail Wood was a female bluebird that nested here in 1964. That was the year of our great Bluebird Summer, the year when, for a few months, hope revived for this tragically depleted species. To a friend, in the season when his father died, Henry Thoreau wrote: "The bluebirds come again, as does the spring, but it does not find the same mortals to greet it. Perhaps there will be a time when the bluebirds themselves will not return any more." During recent decades, the loss of this favorite songbird, once so abundant in New England, this bird so soft of voice, so companionable, so ethereally blue in color, so gentle and appealing in its ways, has become more and more a possibility.

But during that nesting season of 1964 our Hampton bluebirds forged ahead. I know of more than twenty broods that were successfully hatched and reared that year in and close to the village. And at Trail Wood, a single pair produced an unprecedented three broods, one with five young, one with four, and one with two—a total of eleven new bluebirds that our farm added to the Hampton list. I have no doubt that part of this success resulted from the excessive secretiveness of this shyest of females. She was always on guard, always stealthy in her movements, slipping swiftly from the nesting box and disappearing, waiting until all was clear before returning again. The male bluebird would fly directly in with food for the young. But the female took no chances. Even when I was at a distance, if she saw I was looking in that direction, she would avoid the nest. She would perch on a twig or on the top wire of a fence, holding food in her bill, waiting for me to go away. Only then would she make a quick visit to the place where her young were hidden.

The voices of certain birds seem to mark epochs of the year: the drawl of the pewee in hot summer woods, the clamor of Canada geese flying south in autumn, the bell-like trilling of juncoes on the snowfields of winter, the soft melodious warble of bluebirds in the melting days of early spring. When we first hear that gentle sound some of the hardness of winter softens within us. We are always surprised at how far it carries. Standing in the spring sunshine on the slope of Juniper Hill, we often catch this strain, so simple, so rich, coming from the apple trees on the other side of the house.

Vastly different in quality, the voice of another bird clad in blue carries much farther, probably the farthest of any bird of its size we have. Loud, harsh, strident, it cuts through the air, a voice of authority. Countless creatures pay special attention when the bluejay screams a warning. In northern Maine, one summer day, I was observing with my field glasses six loons as they floated far out on a forest lake. A jay called loudly from the shore. On the instant I saw the heads of all six loons turned in its direction. Beside our own pond at Trail Wood, Nellie and I were watching a muskrat swimming out from Azalea Shore in the July dusk when a jay, passing above the water, uttered its screech of alarm. As though in synchronization, the muskrat plunged and disappeared. We have noticed how rabbits crouch down in the grass and gray squirrels sit up and look around when a bluejay calls. Its warnings, although frequently false alarms, are not ignored.

I think the most dramatic instance of the kind occurred early one spring when more than a hundred returning male redwings were filling the treetops along Hampton Brook with the tumult of their calling. The uproar was deafening. Yet when bluejays began to scream the trees instantly fell silent. Not a redwing called. A minute or two passed before the excited chorus rose in volume once more.

One June dawn I awoke early, a little after four o'clock in the morning. As I lay listening to the indefatigable repetition of the phoebe's song outside the middle shed, I began to count the calls. Between four-thirty and five o'clock, the bird repeated its loud emphatic "Phoebe" 958 times—an average of almost one call every two seconds for thirty minutes. In emphasis, phrasing, rapidity of repetition, tonal quality, how varied are the voices of our Trail Wood birds!

We hear the spring song of the white-breasted nuthatch—a short whistled "Whit! Whit! Whit!" like calling up a dog; the goldfinches tuning up their little violins among the thistleheads; the tree sparrows tinkling like a chime of tiny icicles before a winter storm; the broad-winged hawks with their high thin whistled calls that bring to mind "piping the admiral aboard." There is the "Tink! Tink!"—like tapping on metal—of young hairy woodpeckers just out of the nest; the sharp hiss, resembling the hiss of a snake, produced by a brooding chickadee when disturbed within its nesting cavity; the incessant chittering sound of baby downy woodpeckers and a low buzzing that suggests electric wires on a wet night that is the sound of small tree swallows on the nest.

A puzzling occurrence in connection with bird sounds took place on the edge of the village, one April day, a few years ago. While walking near the flooded lowland of Brown Hill Pond, several hundred yards from his home, a friend of ours, Bill Stocking, cut a section from a branch of a sapling springing from the roots of an old chestnut and from it fashioned a whistle such as he had made in boyhood. When he blew it, he was amazed to see red-winged blackbirds and grackles flying toward him from all around the pond. They fluttered about him. They alighted in the treetops close by. They kept him company when he moved. More than a hundred birds followed him home as from time to time, he continued to repeat the sound of the whistle. They remained in the trees of his yard much of the afternoon. What had charmed these birds? What special attraction caused them to congregate? The only explanations that suggest themselves to us is that the sound of the chestnut whistle, in some way, resembled a flocking note or a call for assistance or perhaps even the cry of a young bird in distress.

When he wrote *Rasselas*, in the eighteenth century, Samuel Johnson commented on how birds endlessly repeat the same unvarying songs. These old, old songs—as freshly sung as though never sung before—are, in general, repeated year after year. But they are not, as Johnson supposed, unvaryingly the same. Modern electronic recording methods have revealed that, even more than the sharpest ears can detect, birdsong varies with conditions and individuals. More than a hundred variations of the familiar song sparrow's song have been recorded on tape. Included is the soft little autumn song of these sparrows that we begin to hear when September comes.

Perhaps the variations of individuals, both avian and human, account for the many different descriptions of the song of so common a bird as the robin. To John Muir it appeared to be saying: "Fear not! Fear not!" To many, at the end of a rainy day, it calls: "Clear up! Clear up!" An eighty-year-old woman in Massachusetts once wrote to me that what the robin really says is: "Jonathan Gillett, shine your skillet, shine it nice and clean!" In the sunset of April days, it seems to Nellie the robin sings: "Cheerio! Cheerio! Spring is here-eo!"

In our latitude the birdsong of spring and the migration of fall meet in August. The scarlet tanagers are still singing in the woods when the bobolinks start for South America. Then the autumn tide sets in and the familiar companions of our summer days are swept away. Where are the towhees then? Who hears their bright "Chewink"?

Rarely does anyone observe the exact moment of a songbird's departure on migration. Yet at Trail Wood we once witnessed just such a moment of drama. It came in the spring of 1962. All that winter we had been visited daily by a flock of more than fifty flashing black and yellow and white evening grosbeaks. All of them, except two pairs, left us for their northern breeding grounds early in May. These four birds stayed on and on, trilling in the apple trees, feeding on our sunflower seeds. They were still with us when the trees began to bloom—strikingly colored birds moving among the white clouds of the apple blossoms.

We were watching them on the seventeenth of May—the latest we have ever seen these birds—when, after feeding heavily, all four rose together from the tree. Usually they flew from tree to tree. This time they ascended higher and higher. Their flight seemed more definite and purposeful. They headed directly away into the northeast. We watched them as long as we could keep them in sight. They faded from view and we saw them no more that season.

It was our good fortune that, two years before we moved to Trail Wood, a nature group sometimes referred to as "the Bird Walkers" —The Hampton Bird Club—was organized in the village. Edson Stocking, a graduate of the Yale Forestry School and manager of the Pine Acres Tree Farm, now the James L. Goodwin State Forest, was prime mover and for years president of the club. An association of those who respond to what we respond to, who enjoy what we enjoy, it has proved, for a decade and a half, one of the most congenial groups we have ever known. Enthusiasm is shared. When the telephone rings it may be Evelyn Estabrooks reporting an aurora in the northern sky or Vinnie Stocking telling us nineteen Canada geese are heading our way.

In Webster's dictionary bird watchers are defined under two categories: those who observe birds and those who identify birds—in other words, those primarily interested in the lives, habits, and abilities of the birds, and those primarily interested in their names. From the beginning this village bird club has emphasized observation, getting to know the birds, watching their adventures, studying their ways. Careful lists of species seen on field trips are entered in the minutes but the main concern has been that question asked sixteen hundred years ago by the Greek dramatist Aristophanes: "What sort of life is it among the birds?" Even today the science of ornithology can give only an imperfect and fragmentary answer. The question remains a kind of giant jigsaw puzzle, still only partially assembled. Here and there pieces can be added by amateurs who make careful and accurate observations.

Living as we do among the birds, making our "field trips" by stepping outside the door, here at Trail Wood we feel a closer relationship with them. "The birds of the naturalist," John Burroughs wrote in his first book, "can never interest us like the thrush the farm-boy heard singing in the cedars at twilight as he drove the cows to pasture or like the swallow that flew gleefully in the air above him as he picked the stones from the early May meadows." Something of this dawn freshness, this nearer companionship, is one of the finest features of having the birds of this old farm part of our daily lives.

CHAPTER EIGHT

Complications of the Simple Life

In my journal I find the following entry. It was set down during one of our earliest winters at Trail Wood.

"To the waterfall after breakfast. Begin cutting red osiers that are spreading in the shallow pool above the dam. First a wand springs back and hits me on the lip. Then the ice breaks and one foot goes into the water. After being gone ten minutes, I hurry home with a swollen lip and a boot full of muddy water. Nellie wonders what condition I would be in at the end of a full day's work."

Such are the minor mishaps of the country day, the complications of the simple life. Under the heading "The Simple Life" there are two listings in *The Home Book of Quotations*. The first is: "Its Pleasures." The second is: "Its Faults." Among those of us who live in the country one expression greets all such misadventures as mine. If the well runs dry or a tree falls across the lane or the electricity goes off or a mouse rolls nuts across the attic floor at midnight or the weather man on the radio says "There will be frost for those in the outlying areas"—meaning us—the saying is: "That's country living!"

However, in the main, our decade and a half at Trail Wood has been spent roughing it smoothly. The problems have been fewer than we anticipated. An electric pump supplies us with running water. Large-capacity tanks feed fuel oil into our furnace. Our telephone links us with friends and stands ready in an emergency. An electric stove cooks our meals. Even in the severest gales of winter we have rarely had the electricity go off for more than a few hours. The one exception was the great ice storm of December 1973, when we were without current for four days. And then our big fireplace kept us warm, broiled our meat, baked our potatoes, and kept the water in an old-fashioned iron teakettle steaming on the hearth. Sometimes we have been snowbound for a day or two. But what did it matter? I did not have to go anywhere. In the isolation I could work without interruption.

Although we live nearly fifteen miles from the nearest town, we have dairy products delivered at the door and the laundry truck comes once a week. On each package it delivers, in addition to our name and address there is the notation "Pole 968." In this region the light and power company numbers its poles with metal tags. The one nearest the entrance to our place pinpoints our location for a new driver.

One of our first concerns in settling in the country was to let things alone. A friend of ours once moved from Long Island to a picturesque back-road farm in southern New Hampshire, not far from Mount Monadnock. When we visited him in this beautiful setting, we found he had spent most of his first summer trying unsuccessfully to educate the swallows to enter the barn through another door than the one they had been used to for nobody knew how long. At Trail Wood our external changes were few. Aside from extending our wild paths and adding new gravel to the lane, we left our surroundings almost entirely as they were.

It was in the house that the main alterations were made. Planted solidly on the great stone blocks of the basement walls, the building was essentially in sound condition. It had gone through the hurricane that had toppled the church steeple in the village with no more damage than the loss of a few shingles. But various changes and repairs and additions seemed needed. These included a new roof, new gutters, the rebuilding of the upper chimney, new electric wiring, a new furnace in the cellar, bookshelves along my study walls, and a photography darkroom in the basement.

"If you buy a home in the country and fix it up," an experienced friend had said to us, "you will have a lot of fun and spend a lot of money."

As it developed, he was right on both counts. But it was our good fortune, in connection with this work, to have the benefit of the multiple skills of a local man-of-all-trades. As carpenter-plumber-electrician Barney Pawlikowski saved us weeks of time. When he built my darkroom, for example, he first poured the concrete for the raised floor, then completed the carpentry work, then put in plumbing for hot and cold running water, then constructed the Formica-covered cabinets, then put in the air ducts from the furnace to warm the interior and finally installed the electric wiring and equipment. If I had tried to synchronize the work of a separate concrete man, a plumber, a cabinetmaker, a furnace man, and an electrician, I would have been all summer getting that one job done.

Moreover Barney is well versed in the eccentricities of old houses. He is endlessly ingenious in improvising solutions to the problems they present. The simple uncomplicated jobs tend to bore him. He puts them off as long as he can. But he thrives on emergencies. They present a challenge and arouse his interest.

"If you have trouble," he told us, "call me any time, night or day."

And over the years we have learned he meant it. That is a comforting thing to know in the midst of a winter storm. In this connection, I have never been able to understand how it happens that we have had to call for help so often on Sundays and holidays. One of the riddles of country living is why, if things are going to get out of order, they get out of order on Thanksgiving, Christmas, or New Year's.

The old builders of houses such as ours had small tricks of construction that often were more important than they appeared. In their improvisations they knew exactly what they were doing. When I hired a mason to rebuild the upper part of our chimney, specifying that everything should be exactly the same as it had been, the work duplicated all except for a single row of bricks just above the roof. They were laid flush with the other bricks whereas the original chimney had had that one row set out about half an inch. For years afterward in driving storms the roof leaked around the chimney. That single row of projecting bricks apparently had deflected the water in heavy rains.

In several instances the alterations we made advanced in a kind of chain reaction. One thing made another thing necessary. Before we could install our new furnace, we had to put in a cement floor where for a century and a half the basement had had a dirt floor. And in preparing the cement we found the old well on the terrace beside the house was running dry. So we had to dig a new well farther down the slope where the bottom would be below the level of the brook bed and the supply of pure water would be ample. Digging that well, however, proved easier planned than executed.

The only backhoe available at the time was too light for the job. It scooped out soil and rocks until it descended eight or nine feet into the ground. Then—bang! It hit a solid ledge of rock. The operator filled in the hole he had made and moved the machine on its caterpillar treads to another spot. Again out came dirt and rocks. Again—this time at a depth of eleven feet—bang! Another ledge. The operator tried to break through. A pin sheared off. I drove him home for a new pin. Then he ran out of fuel. With new fuel he resumed the futile banging of the metal bucket on the rock. One after the other, two teeth snapped off. Instead of breaking through the ledge, he decided, he was breaking up his machine. He gave up and rolled off down the lane.

Weeks went by before we could get a backhoe large enough and heavy enough to smash its way through the ledge. In the meantime a thin trickle of water built up into a puddle that eventually became a pool two or three feet deep at the bottom of the hole. The natural history of that sunken pool diverted us during these days of waiting.

It was no more than a couple of inches deep when water boatmen appeared on its surface. These winged aquatic insects come to any patch of water, even the smallest, even water at the bottom of a deep hole. Daily a rain of grasshoppers and crickets also descended into the pitfall. They swam about in the water and crawled out on the little ledges of earth left around the pool by the uneven cutting of the backhoe. When I peered over the edge of the pit in the heat of one August noon, I caught the shimmer of dragonfly wings. A green darner was hovering over the water nearly a dozen feet below the surface of the ground.

Larger creatures followed the insects. Three meadow mice tumbled into the hole. When I looked down they were huddled on the dirt ledges just above the level of the rising water. That afternoon, toward sunset, I slid a long two-by-four down on a slant and left it with the lower end resting on one of these ledges. By dawn all the mice were gone. They had run up the two-by-four and disappeared in the meadow. First a pickerel frog, then a green frog, then more green and pickerel frogs, splashed down into the water at the bottom of the unfinished well. Here they were in their natural element, at home in a waterhole where fallen insects provided ample food.

Less happy was the situation of the next creature to blunder into the pit. I found it there after breakfast on August 25. During the darkness a skunk had lost its footing at the edge and had plunged down into the water. I saw it below me, clinging disconsolately to a narrow shelf of earth. When I lowered some food, a bit of bacon, a slice of bread, the remains of a lamb chop, it ate greedily. During the day I slid a second two-by-four down beside the first. This made a wider ascending walk for creatures trapped in the pit. Morning came. I found the prisoner gone. It too had climbed to freedom up the two-by-fours.

It was the second week in September before the big backhoe, a forty-ton monster, came up the lane and, to our relief, safely crossed the stone slab of our ancient bridge. In order to examine carefully the ledge of rock at the bottom of the hole, workmen dropped down a hose, set a gasoline pump in operation, and sucked out all the water. The next day they would smash through the rock, dig down to ample water, place large concrete cylinders in a column one on top of the other to form the casing of the twenty-foot-deep well and dump in a ton of crushed rock around the lower part of the column to provide a natural sponge. When they left late that first afternoon, nearly a dozen trapped frogs squatted or hopped about on the mud at the bottom of the pit. We could imagine their fate the next morning. We could picture them being mangled by the great steel bucket as it began crashing down as work began. I went to bed planning to arise early, put down a ladder, and rescue the frogs. But when I looked down at dawn, every last frog was gone. Their pond had dried up and they had deserted it. Like the mice and the skunk before them, they all had ascended the inclined two-by-fours and had scattered out into the meadows.

A good many of the complications of our simple life at Trail Wood arise, no doubt, from our interest in and our concern for our wild neighbors. We cannot burn a brushpile because a rabbit lives there. We cannot cut down a barberry bush because a field sparrow is nesting in it. One summer a chipmunk dug one of its holes straight down beside the basement wall. Before we stopped it up, in every heavy rain the burrow formed a conduit, carrying water into the cellar.

Another burrow beside a basement on the south side of the village, one year, produced other problems. In the autumn, a woodchuck went into hibernation in a hole outside Barney Pawlikowski's cellar. It had chosen well. The furnace was just on the other side of the wall and the groundhog spent a comfortable winter in warm quarters. As the season advanced a skunk moved in with it. Several times toward spring the woodchuck roused sufficiently to move about and startle the skunk, which fired its powerful scent gun within the burrow. That scent not only seeped through the wall into the basement but it arrived there exactly where the furnace fan was circulating warm air through the house. The same sequence of events followed each time. Barney would awaken in an upstairs bedroom choking with the fumes. He would jump out of bed, fling open a window to air out the room, and then get back in bed again. The next thing he would know he would wake up thinking he was freezing to death. He had gone to sleep with the window open.

Among all the wild creatures that complicate our lives at Trail Wood probably the most charming is the little white-footed mouse. When autumn comes, it tries to better itself by moving indoors. "Never leave an opening for a mouse" is an old country maxim. Aside from gnawing and nest-building, most of the trouble caused in a country house by these small mammals, with their bright eyes, slender tails, large ears, and pure-white feet, is the result of their provident natures. They mean no harm. The damage they do is unintentional. They just store up things for a rainy day.

Nellie once put on a shoe and found the toe was crammed with sunflower seeds. It was the hoard of a white-footed mouse. Slipped between the sheets of stored newspapers, under a cushion on a chair, behind books on a shelf of my library, we have discovered the caches of these provident mice. One winter the contents of a whole bag of Canada mints disappeared in a night. Weeks went by before we found them. Then Nellie, tilting her vacuum sweeper at an unusual angle, heard a rattling sound inside. She shook it. Out fell a mint. One by one she got the others out. Each had to be maneuvered through a small opening from one chamber into another and then out through a second opening as small as the first. When we counted the pile of extracted mints, we found there were seventy-six. They had all been cached away in the course of one night by the labors of a single mouse.

During another winter a $1,000 mouse lived in Hampton. At the Pine Acres Tree Farm, when a large tractor was stored away in autumn, the manifold was accidentally left off. This opened a path for a white-footed mouse to a secret hiding place within the engine. Night after night it stored kernels of corn on the top of one of the pistons in its cylinder. In the spring, when attempts were made to start the engine, the crushed corn had resistance enough to prevent the piston from completing its upward cycle. Nobody could discover the source of the trouble. Only after the massive machine had been transported to Hartford and the motor taken apart was the mouse's cache discovered and removed. The bill for transporting, dismantling, and reassembling the engine came close to $1,000.

In the fall, as soon as the mice begin coming into the house through some small opening we have overlooked, I begin taking them out again. It becomes an autumn-long game. I live-trap the animals and release them beside stone walls in the woods where they will find protection and natural food. As the years have passed I have transported them farther and farther away. At first I took them just down the lane, but as the number of mice in the house did not seem to decrease, I suspected they were following me home again. And this they may have done. For experiments conducted by government biologists have revealed that white-footed mice have a keen homing sense. Marked individuals are able to return even when they are released at distances as great as a mile away.

In recalling these varied experiences with our wild neighbors, I remember the woodpecker tree. It provided a complicating factor in the days when we were digging our pond.

A pond of our own was the only thing we lacked after our trails were opened and changes in the house had been completed. The most favorable place for digging seemed to be a swampy bowl of land, filled with alders, ferns, and red maples, that lay just down the slope southwest of the house. Springs were numerous there and Whippoorwill Brook, a streamlet originating at the head of a ravine in the woods to the west, added its flow to the swamp. The area of this lowland was about one acre.

Early in May, in 1963, we began cutting and burning the trees and underbrush in preparation for the bulldozer. Everywhere in the wake of the tree cutters we discovered new things. One was our first yellow throat nest, hidden within a green fountainlike tussock of sedge. Another was a spring where, all day long, fragments of mica rose and fell, glittering as they caught the sun, tossed up by the transparent water. But it was a distraction of an entirely different kind that produced the longest-lasting consequences.

On the thirteenth day of this clearing operation, I was standing close by when a power saw sliced through a clump of alders. Unseen at its center grew a poison sumac. In an instant the high-speed chain of the saw's whirling teeth sprayed me from head to foot with its juice. By the next morning the violent burning and itching began. All my past encounters with poison ivy were trivial by comparison. This seemed an ordeal by fire and it continued day after day. With my eyes almost swelled shut, with fiery blisters massed over much of my body, I lay low, rubbing on the latest medicated ointment carried by drugstores as well as the most ancient remedy of all, the one employed by the Indians before the Pilgrims came to New England, the juice of the jewelweed. Nearly two weeks went by before the torment subsided.

In this history was repeating itself. For sixty-eight years before, John Burroughs, on the banks of the Hudson, had viewed the world through one eye for a time while the other was swelled shut as a result of encountering poison sumac. At the time he was helping clear a muck swamp near West Park in preparation for building Slabsides, his famous rustic cabin in the woods.

I was still recovering from my siege of poison sumac when the bulldozer arrived. And this brings us back to the woodpecker tree.

In the course of cutting down the red maples, we had discovered one where a pair of downy woodpeckers were nesting in a hole in a dead limb twenty or twenty-five feet above the ground. We left this tree standing. The parent birds came and went, feeding the young, undeterred by the snarl of the power saws and the crash of trees around them. At last everything was felled except this one tree. We expected the young birds to take wing each day. Through our field glasses we could see them crowding at the opening of the hole, peering down and around, seeing for the first time the world into which they had been hatched. We could hear their high buzzy trills as they grew more noisy in the nest.

But they were still being fed when the bulldozer commenced its work. Its earthshaking bellow passed and repassed beneath their nesting tree. It gouged away the earth on all sides, leaving the tree rising from a tiny island in a sea of mud, a little foothold amid the rawness of the excavation. It was not until the twenty-fourth of June that the young woodpeckers abandoned the tree. As soon as we were sure the nest was vacant, the bulldozer undercut the roots on one side of the island, then pushed against the trunk and rode the tree down in a crashing fall.

In action this yellow mechanical monster, the largest bulldozer in the region, suggested a twenty-three-ton bull of steel bellowing and pawing the ground. The earth shook as it labored up the opposite slope and over the top of Lichen Ridge to deposit more than a truckload of excavated dirt each time in a swampy depression beyond. More than 200 yards away, as I leaned against the frame of the front door looking out, I could feel tremors of the man-made earthquake it produced vibrating through the fibers of the wood.

These vibrations, increasing and decreasing, continued from morning until night. Unexpected side effects resulted from the cumulative shaking of this miniature quake. One whole end of our terrace wall fell down. Apparently the vibrations gradually had jiggled loose some of the anchoring rocks. Next we discovered that a swift's nest had fallen into the fireplace after being shaken free from its hold among the sooty bricks of the upper chimney. In various other ways the lives of different creatures were affected by the noise and confusion attendant on making the pond. Only about fifty feet from the path followed by the bulldozer as it labored up the slope of the ridge kingbirds successfully raised a brood in a wild apple tree. Even more astonishing was the constancy of a pair of towhees.

Their cup nest of twigs, grass, and strips of bark was tucked into the base of a tussock overarched by a low juniper bush. It was no more than half a dozen steps from the raw gash cut in the earth by the machine that, enveloped in thunder, passed so close as it struggled up the slope hour after hour and day after day. There the female brooded her eggs on trembling ground. There the baby birds hatched into a world of overwhelming noise. In spite of these disadvantages, the naked nestlings became fledglings. In time I found the nest empty. The brood had been brought off unharmed in the midst of these abnormal surroundings.

It was on the morning of the day before the Fourth of July that digging ended. The earth stopped trembling and quiet returned to Trail Wood. The raw excavation that lay outspread suggested early pictures I had seen of the construction of the Panama Canal. We could visualize the coves, the shelving shallows, the winding shoreline when the pond was full. Its greatest depth would be eleven feet, its average about six feet, its extent approximately an acre.

On the opposite slope of the basin, I arranged an ascending series of light-colored stones, like the white lines on the hull of a freighter. Day by day we watched the almost imperceptible rise of the water welling from the springs. It gradually reached and covered successive markers. All through the autumn and after winter came we also kept careful watch of the runoff channel we call Stepping Stone Brook. It cuts around the southern end of the low dam to carry away overflow and maintain the level of the pond. In one of those odd coincidences—apparently without meaning but always catching our attention—it was precisely on the morning of New Year's Day when the first trickle ran along its bed, announcing that the pond was full.

CHAPTER NINE

Hidden Pond

That first summer of our pond we seemed living in the Fifth Day of Creation. Life multiplied around us. New forms appeared beside, beneath, on the surface of, and above the water. Using a shovel and a grub hoe, I leveled a circular path at the pond's edge. From it we noted the changes day by day: the arrival of the multiformed aquatic insects—water striders, backswimmers, whirligig beetles; the coming of the bullfrogs; the flight of the swallows at evening, swooping down to drink on the wing; the first tentative green of the waterweeds. Watching our pond was like acquiring a whole new shelf of books, published by nature to be read in the out-of-doors.

About this secluded acre of water, this hidden pond surrounded by woods and hill slopes, a pond you come upon suddenly, unexpectedly as you ascend the lane, there is a special private quality. It is our own companionable pond. Viewing it from our windows, from the path that encircles it, from the little rustic summerhouse above its western end, we have developed an intimate relationship with it.

So today, once more, after a thousand previous times during the past decade, we begin a slow circuit of the path beside the water. We note what we can see and remember what we have seen before. Almost every step along the way recalls events of interest that we have observed from each successive point of view. Nellie's advance is always at a snail's pace. I once timed her on a solitary circuit of the pond and found that, at that speed, it would take her eleven hours to travel a mile. Her motto, she says, is: Go Slow and See More.

Our starting point, this morning, is at Cattail Corner where the trail down the slope of Firefly Meadow joins the path around the pond. Before the coming of the muskrats, an extensive stand of cattails took root here. Redwings nested among their upright stalks. During the autumn migration, one year, an American bittern stayed here for several days. But it is not of birds but of insects that we are thinking as we stand beside the few remaining cattails. Our eyes follow the circles and zigzags traced on the water by half a dozen whirligig beetles. Their movements bring vividly to mind the great whirligig year of 1964.

Each year since water filled the pond, conditions have altered. During the earlier seasons, especially, this new body of water represented nature out of balance. First one thing, then another, forged ahead unchecked. One dawn I remember finding more than 200 newly emerged greenish dragonflies scattered across the dam, clinging to the dewy heads of the clover. Their shed skins were anchored all along the waterweeds of the adjacent shore. Another season clouds of mayflies swarmed up from the water and on a third occasion drifting, dancing curtains of flying gnats, a great mating hour of these insects, left the minute tannish husks of tens of thousands of larval skins littering the pond from end to end. We remember a Summer of the Dragonflies, a Summer of the Bluegills, a Summer of the Whirligig Beetles.

During the latter season these dark-brown aquatic insects, each about the size and shape of half of a coffee bean, rounded side up, flat side down, massed in thousands among the cattails. There they remained during the day. When evening approached, they spread out, following their gyrating paths over the surface of the water. Once in a quick movement I swept my hand through one mass and captured several. For an hour afterward I found my palm was perfumed by a faint delicious odor like that of fragrant apples. Whenever some alarm sent the multitude in spinning motion among the cattails, I noticed that I could detect a slight sizzling sound like fine rain falling on the pond.

In the calm of sunset, when all the beetles were whirling in their feeding, the surface of the pond, seen against the light, was shot with flashing sparks of the reflected rays. At such times tree and barn swallows, skimming just above the surface, cleared long swaths on the water as they sent the alarmed insects gyrating away on either hand.

During this particular summer—the only one in which we had so great a population of whirligig beetles—I watched a large bubble rise through the water, reach the surface, and break in the midst of a cluster of the insects. The cluster disintegrated; the whirligigs exploded away in all directions. Such bubbles, rising from decomposing material on the bed of our pond, provide a kind of natural barometer. We find they are most frequent when the barometer is falling and the pressure of the air is lessening.

Beyond the cattails, the edge of the dam that forms the eastern end of the pond originally ran in a ruler-straight line. It was the pride and joy of the bulldozer operator. For a couple of years after it was finished, I gnawed away at it in my spare time with a mattock, shovel, and wheelbarrow. Now the path winds around little bays and extends out onto small projections. Following this meandering portion of the trail, we recall small adventures of other days. Here on an August evening we saw a bullfrog floating in the water with a damselfly alighted on its back. When the frog swam, the insect rode along. When it drifted without motion, the damselfly sometimes took wing, fluttered out, caught a small insect, and returned to its frog-back landing field. Each evening for several weeks this frog floated in almost exactly the same place in the pond. As we were watching it on another evening, we saw it give a sudden little jump, up and over and down. A turtle had come to the surface directly beneath it.

Approaching the rustic bridge at the far end, the southern end, of the dam, we come to a little point projecting out into the water. I was once standing at its tip when a small red-banded northern water snake swam by. It saw me and dived to the bottom of the pond. There it lay still and I saw its tongue flicking in and out. Apparently, as a snake does on land, it was smelling with its tongue underwater. On two occasions butterflies passing this same point left lasting impressions. One, a monarch, swung low and dragged its feet in the water, leaving a wake of ripples behind it. The other, a spicebush swallowtail, dipped and struck the water with a splash, fluttered on fifteen or twenty feet, dipped and struck again. The butterfly seemed, like the tree and barn swallows, to be drinking on the wing. I have since discovered that there are a number of records in entomological literature of butterflies and sphinx moths that have been seen swooping down in flight to obtain water from the smooth surface of a pond.

Years of weathering have polished and silvered the wood of the rustic bridge across Stepping Stone Brook. We lean on the smooth barkless rails and gaze down into the shallows, into a water scene where something always is happening. Schools of dark little minnows stream past below us. Bluegills patrol around and around above their hollowed-out nests. Beautifully streamlined large-mouthed bass weave in and out among the waterweeds. One drifting nearby opens its mouth in a great O, a prodigious underwater yawn. Another slips among the twigs of a dead submerged tree branch where filmy curtains of green algae wave gently as it goes by. Some days we find scales glued to the bridge railing. There a kingfisher has landed to pound a captured fish on the wood.

I think it was during the second spring of our pond-watching that I lowered the level of the water slightly by deepening the channel of the little brook that carries the overflow 100 feet before dropping it in a cascade into dark swamp woods. A week or two later the bottom of the channel, where the current had been speeded up, was paved with black. I looked closely and found that attached to every small stone were innumerable larvae of blackflies, thousands and tens of thousands of them covering the bed of the little stream. In their larval state these biting flies require more oxygen than most other aquatic insects. They are found oftenest in swift brooks with tumbling cascades where their bodies cover the deluged rocks like dense moss. My temporary change in the brook bed had speeded up the current sufficiently to meet their needs. In succeeding years, when the flow has been more sluggish, we have never seen a blackfly larva in the stream.

Our pond, lying just below the house, is our great weathervane. It instantly records every shift in wind direction. Even the slightest breeze draws along its surface a drifting sheen that seems as insubstantial as an aurora. After days when the wind has been blowing from the northwest, we find flotsam of the pond lodged in the shallows by the bridge. In May it is the white surf of apple-blossom petals arriving from the tree on the northern edge of the water. Or it may be floating rafts of the tannish-red maple keys. In midsummer it is the shed feathers of the mallard ducks going into eclipse plumages. One time, after a vast dispersal flight of ants, the shallows were covered with the lifeless bodies of winged individuals that had descended into the pond. And in October it is the red and yellow autumn leaves that collect in the narrowing mouth of the runoff stream.

One winter, during an idle hour, using a formula based on area and average depth, I calculated the weight of the water in our pond. The figure I arrived at was 16,272,640 pounds. A rise of a single inch in the pond's level adds more than 113 tons, 226,500 pounds. Yet we are amazed each autumn, as the leaves drift down into the entrance of Stepping Stone Brook, to see the level of the pond rise. So frail a barrier as this dam of individually fragile leaves is sufficiently strong to hold back the water temporarily. For days it lifts the level across the whole acre of our pond.

A few yards beyond the bridge two gray glacial boulders, Twin Rocks, rise above the water side by side beneath an overarching red maple tree. There the path swings abruptly westward to keep company with the southern edge of the pond. Rounding Driftwood Cove, where we sometimes watch bass leaping for mayflies in the quiet air of summer evenings, it brings us to a low bench sheltered under a wild apple tree. An ancient grapevine winds among its boughs and, in September, crows and foxes come to get the fruit. Sitting motionless on this bench, partially hidden from view, we are in one of our favorite watching places beside the pond.

As we look out on this occasion we see a kingbird flutter up over the other side of the water and snap an insect from the air. The sharp sound of its closing bill carries to our ears like the crack of a miniature rifle. We watch a small green forest caterpillar lower itself on a silken thread. For a time it swings in the light breeze just above the water. Then it lowers itself a fraction of an inch more and touches the surface. There is a swirl; the caterpillar disappears and a bluegill sunfish swims away. Once, after sunset, looking this way as we ate a picnic supper in the screened-in rustic summerhouse at the other end of the pond, we followed the movements of a great blue heron as it waded out from near this bench. It sank lower and lower until it was beyond its depth. Then it swam for twenty feet or more across a channel to reach shallower water beside the dam.

All around the bench the ground, with its tree roots, is spongy, undercut by the tunneling of muskrats. We watch these aquatic animals, during warm twilights in summer, bringing home food, swimming back and forth across the pond, trailing long ripples, silver V's that grow more silver as the water becomes more dark and mirrorlike. Sometimes they return with mouthfuls of yellow hop clover, sometimes towing long rushes, sometimes transporting such masses of grass stems they have the appearance of green-quilled porcupines moving through the water. Once when I checked on what they were getting near Cattail Corner, I was amazed to find they were harvesting scouring rushes, horsetails so filled with silica they were used by pioneers to scour pots and pans.

On a morning in October, when a light mist hung over the pond, a mink appeared following this path beside the water's edge. It ran in little spurts this way and that, alert, intense, tracing a weaving trail, turning aside, disappearing, reappearing, plunging into the water, swimming swiftly in spite of its lack of webbed feet, climbing rapidly back on land. And all the half hour it remained at the pond, our muskrats swam about without pausing. They circled. They dived. They reappeared again. They kept in constant motion. They made no effort to hide. Their greatest danger was being trapped within their burrows by the mink.

A smaller relative of the mink, a long-tailed weasel, made its appearance beneath the apple tree one day when Nellie was sitting motionless, absorbed in the activity of water striders. In its every movement a weasel gives the impression of intense nervous energy. It is always keyed up. It is like a spring wound almost to the breaking point. Nosing about in small hurried leaps, it approached within a few feet of the bench before it became aware of Nellie's presence. Its back seemed to break in two as the front half reared upward, lifting high its head with its small rounded ears and its dark glittering eyes. For some time it too remained completely motionless. Then, in one swift movement, it vanished in the underbrush to make a stealthy circuit around the bench and reappear beyond to continue its hunting beside Driftwood Cove.

All along the rest of this southern side of the pond we are enveloped, in the early days of summer, in the heavy sweetness of a carnationlike perfume. It is the scent of the white-flowered swamp azalea, or wild honeysuckle. Here too, a few weeks earlier, in May, blooms the pink azalea, or pinxter flower. In consequence, our name for this low lush edge of the pond, bordered by shelving shallows, is Azalea Shore. Beside the path, at one point, there grows a handsome water-loving moss often associated with waterfalls. It has the easily remembered scientific name of *Philonotis*, the pronunciation of which suggests "File-a-Notice."

When fish were few, the shallows beside Azalea Shore were the haunt of innumerable aquatic insects with odd forms and surprising habits. We would stop to watch the slender elongated water scorpions moving in slow motion among submerged plants like walking sticks that had exchanged the land for water. We would observe the aerial dance of the mating caddis flies and the stalking of tadpoles among the waterweeds by the lizard-shaped predatory larvae of the *Dytiscus* beetles. We would see flat slow-moving male waterbugs carrying on their backs 100 or so eggs cemented there by the females. And always there were the adventures of those long-legged skaters on the surface film, the water striders.

We noted how easily they evaded the rushes of sunfish and bass, leaped nimbly over the backs of whirligig beetles, and skated away in spurts of speed when large drops of water, dripping from the leaves, splashed into the pond beside them. Whenever we tossed bits of weed stems out from shore, the water striders, sensitive to the slightest tremor disturbing the surface film, would converge upon them.

One of the strangest recent discoveries in natural history concerns the lives of these insects. In flooded gravel pits in Germany, a few years ago, two scientists studied the habits of a small prey of the water striders, a minute aquatic beetle with the scientific name of *Stenodus*. It is only about a quarter of an inch in length. Yet it employs an amazingly sophisticated chemical defense against its larger and swifter enemy. Within the body of this beetle are two small glands that produce a liquid detergent. When attacked from the rear by a water strider, the insect releases this glandular product. The initial effect is to produce a small wave that propels it ahead in a quick spurt, sometimes at a speed of two and a half feet a second—120 times the length of the beetle. This wave continues for as far as forty-five feet. At the same time the detergent destroys the surface film behind the beetle. Its pursuer, with no surface tension to support it, plunges into the water and drowns. This defense is effective, however, only when the attack comes from the rear. If the water strider approaches either from the front or side, the beetle is invariably caught and killed.

In the earliest days of the pond we sometimes referred to the lowland along Azalea Shore as the Mosquito Coast. To keep down the mosquito larvae, we introduced the beautiful little golden dace. That was the beginning of another unbalancing of the pond. The dace, freed from normal predators, multiplied prodigiously. To keep down the dace, we introduced bluegill sunfish and then, in a misguided moment, added half a dozen bullheads. Each new addition erupted into a population explosion of its own. Finally we introduced the voracious large-mouthed bass. Dace and bluegills and bullheads grew fewer and fewer. A balance of sorts has been produced by the bass. They are the climax fish of the pond.

As we circle the edge of Whippoorwill Cove and pass below the little cataract where Whippoorwill Brook drops into the pond and turn to the north to follow the western shore, a convoy of bass keeps us company. A dozen or so, like a flotilla of submarines, cruise beside us. On occasions bass and bluegills follow us halfway around the pond. Their interest in us is like that of the swallows that swoop about us when we walk through meadow grass. It is the insects, the leafhoppers, the crickets, the grasshoppers, that we frighten from the weeds beside the path—some of which alight on the water—that is the attraction for the fish. We are their unwitting benefactors.

Always on the water of the cove insects are floating. Among them are commonly robber flies lying on their backs, their slender legs in the air. These swift aerial hunters overtake their prey in sudden spurts of speed. They are fast; but they lack endurance and in flying across a body of water they soon tire. They sink lower and lower; their feet touch the water; they flip over on their backs and are trapped. Yellowjacket wasps are the only insects I have seen that have been able to save themselves after they have landed on their backs in this manner. They somersault over onto their feet again and take off from the surface film. This they do in one swift continuous movement. On the hottest days of summer, at the edge of this sheltered cove, these wasps often come to drink. I see them alighting on the shallow water, drifting for a while, then taking wing again. When a breeze is blowing, the floating insects sometimes go drifting by like tiny sailboats.

Most of the insects—flies, beetles, small white moths—that are trapped on the water struggle for a time, often in the midst of brilliant shimmering pinwheels of reflected light. But they usually turn this way, then that, circle in one direction, then in the other. So they exhaust themselves, frequently within a few feet of shore. In contrast was the effort of one small red and black ladybird beetle. When we first caught sight of it, it was a couple of feet out in the water, swimming steadily with its six hairlike legs, heading directly for land. It kept coming, following an almost straight line. Slowly, minute by minute, it narrowed the gap to safety. In the end its straight-line course carried it to a small rock at the pond's edge. We saw it crawl up this and disappear. It had escaped from the perils of the water unharmed.

Summerhouse Rock, one of seven glacial boulders scattered around the margin of the pond, marks the end of Whippoorwill Cove. Its flat yellow-gray surface, eight by eight and a half feet, projects out into shallow water. Near it robins, catbirds, redwings, grackles, thrushes, and scarlet tanagers have a bathing place. Almost from the other end of the pond we sometimes can hear their splashing.

Standing at the outer edge of this platform of stone, we gaze directly down on all the submerged activity in the water below us. We watch fish drifting by with a slow, lazy flicking of their tails. We see hundreds of alarmed tadpoles race away, each leaving a trail of stirred-up mud along the bottom. But in all the thousands of times I have stood upon this rock watching the pond life around and below me, the strangest thing I saw occurred one morning in mid-July. It provided a momentary glimpse into the bizarre and incredibly keen senses possessed by creatures that live beneath the water.

The nests of the bullheads, at that time, had produced their teeming masses of minnows. One such mass, which may well have contained 1,000 little fish, was gyrating slowly over the mud of the shallows near the shore, five or six feet away. Like a black spiral nebula, with trailing arms extending out around the edges, it continually revolved about its own center. Two parent bullheads swam back and forth on guard. Whenever they were frightened away, I noticed how, with a sweep of their tails as they left, they sent up a cloud of mud that drew a concealing curtain of sediment over the whirling mass of their offspring.

After I had been watching them for some time, I tried an experiment. I waved my arms in the air. The small fish paid no attention. The motion did not disturb them. I produced a sharp sound by tapping two stones together underwater. There was no response. Then I stamped my foot on the great stone on which I was standing—a stone that surely weighed half a ton. The result was instantaneous and explosive. The mass of black minnows seemed blasted apart. Fragments shot in all directions. Then the school regrouped itself and commenced revolving once more in its feeding.

How incredibly sensitive to vibrations these minnows must be! On later days I repeated the experiment many times, sometimes on the flat rock, sometimes on the softer surface of the path along the pond edge. In each instance when I stamped down my foot—even when I was as much as twenty feet from the minnows—the slight tremors I produced were sufficient to trigger, like a silent explosion, an instantaneous reaction that tore apart the mass of baby bullheads.

How did the fry detect such infinitesimal vibrations? What ultrasensitive organs carried the warning of danger to each at the same split second? Fish that swim in dense schools have been found to possess delicate pressure organs along their sides that warn when they come near other objects. Thus they are able to turn and twist and circle as a unit, each component fish keeping its distance from its neighbors. Perhaps it is such pressure organs that catch the unfamiliar tremors and warn the inexperienced fish of danger.

During some summers, all along the curving edge of the pond from Summerhouse Rock, past the little mud delta produced by the wet-weather flow of Woodcock Brook, and on to Turtle Rock, the warm shallows seem paved with pollywogs. By taking sample countings, one year, we estimated our pond contained 10,000 tadpoles. So great is the allowance nature makes for death and destruction! Besides the American toad, *Bufo americanus*, six kinds of frogs lay their eggs and produce their tadpoles in the Trail Wood pond—the hyla, or spring peeper, *Hyla crucifer*; the bullfrog, *Rana catesbeiana*; the gray tree frog, *Hyla versicolor*; the green frog, *Rana clamitans melanota*; the pickerel frog, *Rana palustris*; and, more rarely, the leopard frog, *Rana pipiens pipiens*. When, on spring mornings, I circle the water before breakfast, I find the gelatinous strings and clumps and sheets of eggs that have been produced in the night spreading out among the waterweeds. Half a dozen times each year I discover among them imprisoned dragonflies, their wings held fast by the batrachian glue of the egg masses.

A few days after these eggs first appear, we watch the tiny frogs-to-be hatching, dropping to the bottom, littering the mud like small black elongated seeds. Later still, now with their bodies swelled like little balloons, dark above, light below, with slender tails lashing rapidly to drive them through the water, we see them swarming amid submerged clouds of green algae. There they find both green pastures for feeding and a place of refuge from their enemies. Bullfrog tadpoles, later in their lives as their diet widens, become jumping tadpoles. We see them breaking the surface of the water like leaping fish, snapping up, with a small but audible sound, small insects floating there. Later still we watch them acting as underwater scavangers. Once I saw half a dozen of them, their tails extending out like the points of a star, working at the body of a drowned mouse.

Whether they are bullfrog tadpoles that may remain tadpoles for three years or little scarlet-tailed tadpoles that hatch from eggs cemented to submerged grassblades, develop into translucently green froglets and eventually become mottled lichen-colored gray tree frogs, all these underwater creatures undergo the same great and dramatic change. In season we see them all around the pond swimming with tiny legs hanging down. From gills to lungs, from fish tails to hopping legs, from a world of water to a world of air, their transformation pursues its course. This is the time of the almost-frogs. This is the period of breaking down, absorbing, reforming, a time of alteration almost as complete as that experienced by a larva when, within its cocoon, it transforms into a moth. During these days, when their mouth parts are changing, tadpoles cease to eat. Then suddenly, everywhere along the margins of the pond where only yesterday it seemed there were tadpoles, today there are small frogs leaping. Crows find rich hunting when they alight along the path. Grackles bring their broods, just off the nest, to dine on the swarming froglets.

Those that escape crow and grackle and heron, water snake and bass, may live through a number of seasons. The life span of a gray tree frog may extend for as much as seven years; that of a bullfrog for as much as sixteen. So year after year we may hear the voices of the same individuals raised in the dusk of the spring nights. They vary from the frail and lonely twilight sound of the spring peeper to the low-pitched twang of the green frog to the small snoring sounds of the leopard frog and the fluttering musical trill of the gray tree frog. Loudest of all is that foghorn voice of untold thousands of muddy ponds across the country, the deep "Ter-rumpp" of the yellow-throated male bullfrog. More than once we have caught a faint low "Gur, Gur" coming from the water. It has been the voice of the pickerel frog. This grating sound is often produced when the frog is submerged and resting on the bottom of the pond. In the midst of this spring chorus of the frogs there rises the most beautiful of all our batrachian sounds, a pure sustained trill that goes on and on. It is the mating-time music of the American toad.

When we leave behind Turtle Rock—that favorite sunning place of the painted turtles—the path swings abruptly to the right and we find ourselves following the fourth and final side of the pond, the north side. Near the next landmark at the water's edge, Beaver Rock, memories come crowding back. This rock, another relic of the glaciers, when seen from the viewpoint of Driftwood Cove, resembles a giant beaver coming to shore. It was here we watched a fisher spider descending to hunt underwater, here we observed pickerel frogs heading out in the dusk for a night's hunting in the meadows, here we saw a mallard duck break off and progressively swallow a piece of cattail leaf a foot or more in length. And here, one summer, we discovered that the sides of the rock and all the pond bottom around it were densely littered with the mottled shells of the little tadpole snail—so called from its slender, pointed foot that suggests the tail of a tadpole.

The Duckling Path, along which a mallard female one year brought its brood from a nest hidden in the tangles of Veery Lane, ended beside Beaver Rock. Under it, that year, there lurked a twenty-pound snapping turtle that had wandered up Stepping Stone Brook into the pond. It destroyed many of the ducklings before it, in turn, was destroyed. Not far from this rock, for several weeks during another summer, we observed a bullfrog that was blind in one eye. When something alarmed it, we would see it whirl in a swift circle, sweeping its good eye in a complete 360-degree circuit of its surroundings.

We watch our step when we pass under the wild apple tree a few yards beyond Beaver Rock. Here the ground is spongy, the footing uncertain. Muskrats have undermined the path in their tunneling. It is probably no coincidence that the two places where these animals have burrowed most extensively have been beneath the two wild apple trees that grow beside the pond. The interlacing roots, no doubt, provide their burrows with additional strengthening and protection, making it more difficult for an enemy to dig them out.

One afternoon in early October, we were looking down this edge of the pond from farther up the path when we saw an animal, its head much smaller than a muskrat's, swimming away from shore near the apple tree. It made rapid progress with long ripples spreading away behind it. It was heading for Azalea Shore, starting to cross the pond at its widest point. I swung my glasses in its direction. The little face was striped. The swimmer was a chipmunk.

In one of the volumes of his *Lives of Game Animals*, Ernest Thompson Seton tells of a chipmunk, closely pursued by a weasel, that took to the water with its deadly enemy just behind it. For a hundred yards this life-and-death swimming contest continued. Steadily the chipmunk drew away from its pursuer and escaped. Luckily for our chipmunk, none of the large-mouthed bass saw it. It had almost reached the middle of the pond when I waved an arm and called:

"Hey, chipmunk! You better turn around and go back!"

Probably my movement, rather than my advice, was responsible, but immediately it reversed its course and headed back for shore. Running along the path, I was close to the spot when it emerged. Its fur was unmatted. It was almost dry. It had shed water like the feathers of a duck.

The path beyond the apple tree runs straight to Cattail Corner past a great glacial rock and carries us back to our starting point.

It is from this northern side of the pond, looking down the slope from the house, that we observe those two landmark events in the year of the water, the coming and the going of the ice. By early December, we see the first skim around the edges. Sub-zero nights follow and the hard, cold lid of the solid ice clamps over the pond. Lying outspread in a snow-covered expanse, it hides all below it while the winter passes by. Then, in lengthening days, the retreat begins. A dark rim of open water widens around the edges. The unanchored raft of the central ice shrinks, becomes more spongy under a higher sun. Usually on some latter day in March we watch the last remnant melt away. Now the eye of our pond is open. The life of the water awakens. And, for a whole new season of warmth, we are free once more to enjoy all the little adventures that lie along this bordering path of ours.

CHAPTER TEN

The Starfield

Those who leave the city behind and move to the country return not only to the open fields but to the open sky as well. Stars and planets and constellations become companions of the country night. In the clearer air over the darkened fields, they seem to draw closer, to burn with greater intensity, to increase into swarming multitudes. No longer do city lights dim their brilliance.

Lying in a reclining chair on clear summer nights, looking upward into the face of the open sky, I enjoy a kind of nocturnal counterpart of my hammock in the woods. Here I watch the heavens as there I watched the treetops. In the darkened sky I see Vega, the star of summer, as Sirius is the star of winter. Over me streams that river of stars, the Milky Way, known to the ancient Chinese as "the Little Sister of the Rainbow." As I lie there, during hours when the atmosphere is unusually limpid, the illusion grows that all the stars and constellations are swinging low, are shining almost within reach of an outstretched arm.

Our interest in the night sky above our fields is, no doubt, more poetic than scientific. It is pure enjoyment rather than serious study. We recognize the constellations, the Swan, the Lion, Cassiopeia and Andromeda, Canis Major and Canis Minor, Pegasus, Orion and the Dragon, as acquaintances of ours in the vastness of outer space. We see them move, seasonal and nightly landmarks, across our sky. An hour thus spent under the stars, in the dark of the moon or before its rising, with the heavens alight from rim to rim with the gleam and glitter of planets and constellations and galaxies, is an ethereal experience, a calming prelude to a night of rest.

Our favorite spot for watching the night sky is the open elevated ground of the north pasture. We early named it the Starfield. Particularly here in the cold brilliance of winter nights, when we are overarched by the luminous clouds of the Milky Way galaxy, when Orion comes into its own and the burning stars of the Big Dipper wheel slowly above the black fringe of the treetops to the north, is the celestial show above us at its best. Then we gaze up at the clustering Pleiades that glittered for the English poet, a hundred and more years ago, "like a swarm of fireflies tangled in a silver braid." In December the stars come early. Then, as John Josselyn noted in seventeenth-century New England: "By five of the clock it is pitchie dark." But it has not been only among the frosts and snows of winter that the Starfield has served us well. There we have surveyed the night sky in spring when the oak leaves were yet to become "as large as a mouse's ear"; in summer when mist in the hollows was shot through with the gleams of those tiny stars of the earth, the fireflies; in early autumn when we sometimes heard the rustle of earthworms among fallen leaves in the darkness around us.

At the lower edge of the Starfield, one year, we dumped several loads of fireplace logs. They made convenient seats where I could brace my elbows on my knees as I looked through binoculars or peered into the eyepiece of a compact twenty-power telescope. I recall one particularly brilliant night there. I swept my glasses back and forth across the heavens. I spent a long time at the nebula of Orion. I noticed how the Pleiades held little trailing lines of fainter stars. The night was calm, the weather comparatively mild under the glitter of all those swarming stars—little stars, new stars, stars usually unseen.

When Leslie Peltier, the famous amateur astronomer of Delphos, Ohio, set up a larger telescope on the edge of the Starfield one October night in 1964, we looked at the sky with vision expanded. Although the night was rather hazy, we saw distinctly the rings of Saturn and the craters of the moon. I have often imagined since what might have been revealed, given the limpid atmosphere, the crystalline sky of other nights.

But even under the clearest sky, even with the aid of an observatory telescope, most of the stars remain invisible to us. When, night after night, I look toward the Big Dipper in the north, I see the part of the sky enclosed within its bowl as a dark and empty space. Yet photographs taken through one of the larger observatory telescopes reveal that in that "blank" space there are at least 1,500 galaxies of stars. Harlow Shapley, the Harvard astronomer, has estimated that a careful search of this area with the 200-inch telescope on Mount Palomar would record a million galaxies, each galaxy comprising something like 10 billion stars. Thus that comparatively small celestial area contained within the bowl of the Big Dipper—an area that to my eyes appears almost wholly darkness unrelieved by starlight—may hold, in reality, more than 10 million billion stars.

In earlier times, the night sky was more closely part of the lives of those who farmed our Trail Wood acres. Sighting along the east side of our house after dark, we find it is lined up exactly with the North Star, Polaris. The builders, more than a century and a half ago, apparently insured that the house would face precisely toward the south, as so many houses in the area do, by setting up stakes and lining them up with this stable guidepost amid all the apparent movement of stars and planets around it.

In our winter skies the brightest constellations burn in the southern half of the heavens. There Orion, the "White Tiger" of the Orientals, is the first to catch the eye. On such nights, when I look up into the hickories beside the lane, they appear as star trees. Their maze of bare branches rises in black silhouette as though decorated from top to bottom with the glittering brilliance of the stars shining behind it. The trees represent the present, the stars the past. For, as William T. Olcott observes in his *Fieldbook of the Stars*, for so long have their rays been journeying through space that: "When we scan the nocturnal skies we study ancient history. We do not see the stars as they are but as they were centuries on centuries ago."

In contrast, however, on those August nights in the Starfield when we watch the Perseid meteors consuming themselves as they draw bright, thin lines of fire against the background of the stars, we are seeing an astronomical event instant by instant as it occurs. Each falling star is part of not ancient but present up-to-the-second history. In *The Golden Bough* Sir James G. Frazer recalls an interesting instance of the special importance of meteors in the time of the Spartan kings. A rule of the constitution provided that every eighth year five magistrates should "choose a clear and moonless night and sitting down observe the sky in silence. If during their vigil they saw a meteor or shooting star, they inferred the king had sinned against the deity and they suspended him from his functions."

Other lines than those drawn by meteors—lines brilliant, brief, jagged, and infinitely nearer to us—are a feature of our summer nights and summer days. Lightning is common in our region. It is a rare house that does not carry lightning rods. A local belief is that thunderbolts may be drawn to our particular area by water held near the surface by ledges of rock. But, so far as I can find, during all its years on this knoll above the brook, our house has never been hit directly by a bolt from the sky. Because lightning tends to strike the tallest object in the vicinity, it may obtain some natural protection from the lofty trees just across the lane.

Evidence supporting this idea came one muggy afternoon in late July. I was working at my desk when a summer storm struck. Rain descended in a sudden sheet of water. Then came the lightning. Tremendous bolts struck to the north and west. The air teemed with electricity. Little flashes raced in all directions. Twice there were loud snaps when charges of electricity reached our lightning rods. Gradually this violence subsided and the sky lightened and the rain stopped as the storm moved away toward the east. Then—as in a parting flash; almost the final lightning of the storm—a thunderbolt struck one of the hickory trees thirty-five feet from where I sat. The dazzling glare and the thunderclap exploding like an artillery shell came together. The bolt leaped from place to place as it descended the trunk, blasting bark from the tree and hurling it fifty feet away. It bridged the gap to a fence post, shattered it, and sent the fragments flying. Then, after following the barbed wire of the fence for a rod or so, it plunged into the ground and disappeared.

That thunderbolt probably was a relatively small one. The awesome power of a major stroke was demonstrated the following year some two miles to our west, at the edge of the James L. Goodwin State Forest. Here the bolt struck a telephone pole, crossed a road on an overhead bracing cable, melting it in its passage, and then for 1,000 feet followed off into the forest the wandering line of an ancient barbed-wire fence that had fallen and was buried in the rain-soaked ground.

When we visited the spot a few days later we found a trench plowed through the woods such as might have been torn up by a rampaging torrent of spring. Instantaneous steam generated by the incandescent passage of the lightning bolt had produced explosive pressures. In places the groove among the rocks and trees was a yard deep and nearly as wide. Once we found a stone, three feet square and weighing perhaps 200 pounds, that had been hurled ten feet from the trench. When we looked beneath we saw small sapling trees, holding green leaves, flattened by its fall. Farther on, in another demonstration of the tremendous power unleashed by this single flash from the sky, we noted a larger rock that had been split in two. It may well have weighed 500 pounds. One of the halves had been lifted so high into the air it had knocked bark from a tree trunk three feet above the ground.

We came to a place where a wire of the fence, many years before, had been stapled to a small oak tree. The growth of the tree had enveloped the metal until both wire and staple were buried deep within the trunk. In following the wire, the thunderbolt had split the trunk and bowed out the two halves. Peering into the gap, Nellie and I could see the staple still anchored to the wood of the interior of the tree.

Everywhere along the course of the lightning-dug trench, the lower leaves of the trees had been shredded or stripped away entirely. Debris from the woodland floor had been flung high into the treetops. All around us it hung from twigs and branches like flood wrack or Spanish moss in southern swamps.

Less fleeting than the glare of lightning or the bright traced line of the falling star was another flaming path that cut across our eastern heavens in the spring of 1970. It was the path of the rarest and most spectacular visitor we have seen in the Trail Wood sky. About four o'clock in the morning, on the first day of April, we looked out a bedroom window facing the east. Our attention was attracted instantly by a strange radiant vision glowing in the sky just above the black silhouettes of the trees rising beside the brook. With its brilliantly luminous head and its long trail of gossamer light extending away behind it, Bennett's Comet was dominating that portion of the heavens. Not in more than half a century, not since the famous Halley's Comet of 1910, had one of these heavenly bodies burned with such splendor in the sky above America.

I remember that, as a boy in the Midwest, I stayed up until late at night to watch Halley's Comet. It is the most celebrated of all these visitors from remote space that, in their wide-ranging orbits around the sun, return at regular periods. For more than 2,200 years, ever since 240 B.C., it has been recorded at intervals of seventy-six years. During the spring when we watched Bennett's Comet our eyes often followed the filmy light of its nebulous tail stretching across the sky. We frequently wondered how long it was. We never found out. But we did discover that astronomers have calculated that such trailing wakes of luminous dust and gases have extended behind some of the larger comets for as much as hundreds of millions of miles.

Early in September, one year, Nellie and I were sweeping the weeds and grasses and bushes along the lane with the beams of flashlights, tracing to their sources the songs, shrill or mellow, of nocturnal insects. We had come to the bridge and were standing in darkness, listening to the rush and gurgle of the brook among the stones of its bed, when we noticed a tiny greenish light glowing at the edge of the road. It was the feeble lamp of a glowworm. We bent over it just as, on an evening in May, in 1852, beside a winding road descending from the village of Lancy to Geneva, the Swiss philosopher Henri Amiel had stooped to watch another glowworm, which, as he recorded in his *Journal Intime*, was crawling furtively under the grass "like a timid thought or a dawning talent."

When I looked up from that tiny gleam of living light in the darkness at the edge of the lane, my eye was caught by a movement in the northern sky. In waves and streamers silvery light was sweeping upward among the stars. In an instant our attention shifted from the frail minute spark of the glowworm to the vast shimmering celestial display of our first aurora borealis at Trail Wood.

All above the black line of our North Woods, the sheets of light were growing brighter. They waved in a moving band across the sky, a band that increased in intensity, faded, strengthened again. The streamers paraded from east to west. New shafts continually speared upward while the western sky grew rosy with a deepening reddish glow. Once a falling star drew its fine line of illumination across the northern lights. For nearly an hour the auroral display continued. At last the rosy hue in the west began to pale. Then, as we watched, the streamers, like the lessening flames of a dying fire, sank lower, grew dimmer, gradually faded from the sky.

Other auroras, seen from the Starfield in subsequent years, remain vivid in recollection. In one we had the impression the shimmering light was descending like fine curtains of rain around us. In another, lightning came and went far off in the northwestern sky. In a third, over and over there drifted down to us the small frail voices of unseen songbirds migrating south across the vastness of a sky pulsing with the silvery shimmer of the polar lights.

Each of these magnetic storms is triggered by an explosion on the sun. In the far north, where lines of magnetic force come together, such displays at times are almost continual. Thus, in a sense, the aurora provides a substitute sun that makes less dark the long night of the Arctic winter. In rhythm with the eleven-year cycle of sunspots, auroras increase and decrease in frequency. During the mid-1960s, we were in a period of relative inactivity, "the years of the quiet sun." Then in 1969 the excitement of auroral displays returned. On the twenty-third of March that year the most spectacular aurora of our lives began about 8:30 in the evening.

Bundled up in the thirty-degree temperature, we watched it from the open pasture land of the Starfield. Around us patches of snow in the hollows of the meadow were flushed with a faint reddish tinge reflected from the glow in the sky. The air was still, the night silent. The play of auroral light absorbed us completely. We lived in our sense of sight.

Sometimes silver, sometimes frosty green, drifting or shooting upward, the sheets of shimmering light ascended toward the zenith. In sudden rushes they spread to right or left as though gusts of wind were driving powdery luminous snow before them. Everywhere the play of light was shifting, glowing, waxing, and waning. The whole sky seemed alive. We had the sensation of continually catching fleeting movements from the corners of our eyes.

In the north and east and west radiant patches, at first rose red, expanded and deepened in color. At times they were bordered by light almost electric blue. An airliner, high above us, flew through one reddish patch, its own red lights winking on and off. Toward the west, close beside one of the largest areas of red, a pale crescent moon gleamed faintly. To the north, over the ragged ebony line of the woods, the Big Dipper imperceptibly wheeled on its course through streamers and curtains of silvery light. Once, recalling our first Trail Wood aurora, a falling star drew an almost horizontal line across the east through veils of white and patches of red. Here and there, where the glowing areas and the rising and falling curtains of light grew thin, stars shone brightly as though through windows.

And all around the whole 360 degrees of the horizon, that night, shafts of white, at times suggesting the beams of searchlights, soared upward, frequently seeming, at the top of the sky, to overlap like the poles of a tepee—a vast celestial tepee with all the dimly visible earth enclosed beneath. In the stillness of the night, we had the impression of standing beneath a luminous insubstantial tent, a shimmering canopy of light. Never before nor since have we witnessed an auroral display of such magnitude—extending completely around the horizon. It was seen, during this night of early spring, by those living as far south as Lake Charles, Louisiana.

For nearly two hours we watched while the movement and color continued. So stirring was the experience that we felt lifted from the earth. The beauty of an aurora is all spirit. Even long after, the recollection of it brings back the emotion of magical moments. No other contact with the heavens at night that the Starfield has afforded has been so ethereal, so memorable, so moving as those times when the shifting, pulsing light of an aurora has filled the sky.

CHAPTER ELEVEN

Windows on Wildlife

A cottontail rabbit came hopping over the drifts in the dawn of that third day after Christmas. The thermometer stood at zero. But the nightlong gale, the winds, and the driving snow were gone. We stared through our bedroom windows on a cold white winter world where the tree trunks were plated with snow on their northern sides and sweeping drifts curved up to the top of the walls along the lane. In all that frozen expanse the only animal life we saw was the small dark form of the cottontail rabbit.

Halfway up the lane, between the bridge and the hickory trees, it stopped, turned aside, and ascended the smooth featureless breadth of the drift. There it paused for a moment. It looked intently about for possible enemies. Then it began digging rapidly with its forepaws. Snow streamed out behind it. In a few moments its head, its body, its cotton tail, and finally its long kicking hind legs disappeared as it tunneled in an almost vertical descent through the drift. Less than a minute went by before it backed out and carefully looked around in all directions. Then it disappeared into the hole again and the snow came flying out as before. After this had gone on for several minutes, the rabbit reappeared, cotton tail first. Moving back down the drift about ten inches, it commenced digging once more. This time it vanished into the snow. Although we watched for a long time, it did not become visible again.

Later that morning, I plowed through the drifts to the spot. When I peered into the two holes, I discovered that the first tunnel ended at frozen ground. But the second descended directly into the entrance of a woodchuck burrow. There, on this zero morning, the cottontail huddled snug and warm in a sheltered retreat.

Creatures that save their lives by flight, as rabbits do, are always aware of where they are in relation to places of safety. Their territory sense must be particularly keen. For when a predator appears suddenly there is no time to look around for a hiding place. This we realized. But never before had we seen a rabbit's sense of location so dramatically demonstrated as on this winter morning. Even on that white uniform unmarked expanse of the snowdrift, the cottontail had missed the hidden woodchuck hole by less than a foot on its first attempt and had scored a bull's-eye on its second.

Living as we do in a secluded house in the country, surrounded by woods and fields and streams, we spend the year as in a permanent observation blind. Looking out our windows, we continually see things of interest. Not infrequently we witness occurrences we have never observed before. All the birds and mammals around us are existing under natural conditions. They are not tame; they are not pets. They live in a kind of symbiotic relationship with us. Our presence provides them with a certain amount of protection; their habits and activities provide us with entertainment, instruction, and pleasure. The windows of our house, facing as they do in all directions, are windows on wildlife.

I remember how, for several days one year, we were diverted by the actions of an indomitable gray squirrel. We first noticed it soon after I hung a roofed-over bird feeder on the clothesline attached to the apple tree just northeast of the kitchen window. It had dropped down from a branch above the feeder and while birds waited in the tree it sat stuffing itself with sunflower seeds. I slid the feeder along the rope beyond the end of the branch. When we looked out again the squirrel was where it had been before. It had easily leaped from the tip of the limb. I shifted the feeder farther away from the tree. The squirrel climbed to a higher branch and bridged the gap in a longer leap. This went on day after day while I advanced the feeder farther and farther out along the clothesline. Each time I moved it the squirrel climbed higher into the apple tree. It was like a game of checkers. I made a move. It made a move.

Finally, when the feeder was a dozen feet from the tree, the squirrel landed on it with such force it could no longer cling in place. It hit the hanging feeder a glancing blow that made it bounce and tilt and then it went spinning on through the air, twisting so it landed on its feet in the soft grass below. But it was still unvanquished. It was still ahead of the game. For each time it struck the feeder, sunflower seeds spilled out and showered down on the grass. There the gray squirrel fed until they were gone. Then it went scrambling up the tree to make another leap with the same results. Before it gave up, I had moved the feeder almost fifteen feet out from the tree and the squirrel was climbing to the upper branches before launching itself out into space toward the hanging target below. How far was it leaping? As nearly as I could calculate, the little animal, in its flying jumps from the limb to the feeder to the ground, was traveling almost, twenty-five feet through the air.

Near this same apple tree, on a day when four or five inches of light fluffy snow had fallen, we viewed the antics of another animal, a smaller red squirrel. It would plunge into the snow, tunnel like a high-speed mole beneath the surface and pop up half a dozen feet away. Watching it was like watching a loon diving on a northern lake. We could rarely guess where it would emerge. The little animal, so intense in its every action, appeared to be having fun, playing a game, enjoying itself by sporting under the soft blanket of the new-fallen snow.

In contrast, there was another animal that I watched from my study window one year on the first day of March. At first glance it suggested some Jurassic monster in miniature as it toiled slowly across the ice of the pond. It was an opossum out in the winter weather. Almost at the opposite extreme from the lively red squirrel, this primitive mammal moves as though permanently locked in low gear.

Even slower and even more surprising was another and smaller creature I gazed down upon when I looked out another study window one morning toward the end of December. A black and brown woolly bear caterpillar crawled over the frozen crust of a drift below. I checked the thermometer. The mercury stood at twenty-four degrees F.—eight degrees below the freezing mark. There was no wind that day. The sun was bright. From time to time I saw the creeping caterpillar curl up and lie on its side. It would remain thus for several minutes, like a woolly doughnut on the snow. Its stiff hairs lifted it above the cold frozen surface below and its dark colors absorbed warmth from the sun. It may be that the bodies of these larvae contain some "antifreeze" chemical such as has been found inside carpenter ants during the period of their winter hibernation.

Through various windows at various times we have seen such things as a crow hunting in the yard and pulling earthworms from the ground like a robin. We have seen a white-breasted nuthatch—instead of following the usual procedure and hiding them in crannies and crevices in the bark of trees—secrete sunflower seeds in grass clumps. We have seen a muskrat, in the dead of winter, struggle through the snow up the slope from the pond and feed on cracked corn put out for mourning doves. We have seen red-winged blackbirds, just as roadrunners turn over flat pieces of dried mud in the Southwest, flip over plates of frozen snow in early March to expose the birdseed attached to their undersides.

Through our windows we have noticed how woodchucks carry dry leaves into their burrows for bedding while chipmunks bring in mouthfuls of damp or wet leaves. The reason, I think, lies in the size of the holes. The burrow of a woodchuck is large enough to admit the dry brittle leaves without breaking them while the small entrance of a chipmunk's hole will accommodate only the softened, more flexible dampened leaves.

On the day before Christmas in 1968 a mourning dove feeding on cracked corn scattered along the lane flew up and struck my west study window in a hard glancing blow. It left behind a small fluffy feather attached to the glass by the tip of its shaft. Apparently it was cemented in place by a bit of torn-out skin or flesh that dried and hardened and became as strong as rawhide. Day after day I watched the feather fluttering first in one direction then in the other in the changeable winds. Each morning I expected to find it gone. But it survived sleet. It survived snow. It survived gales. It survived freezing rains. It remained there while I worked on five chapters of *Springtime in Britain*. Week after week I saw it clinging in place whenever I looked through the glass. I wondered each morning how long it would be before it lost its grip. When at last I found it dislodged and frozen in ice at the bottom of the pane, it was the first day of March. Soft wet snow, packing around it, had given the gusts greater purchase and they had torn it free from the glass. But it had remained there all but three weeks and three days of the entire span of winter.

Toward the end of August, one summer, I was working here at my desk when Nellie called from the kitchen:

"Here's something I've never seen before."

I joined her at a window. On the ground beneath a plastic jug we were using as a feeder for chickadees and nuthatches and purple finches, a chipmunk and a white-breasted nuthatch were hopping about, seeking fallen sunflower seeds. Each time the chipmunk neared the nuthatch, the bird spread out both wings to their fullest extent and fanned wide its tail. Then it began slowly rocking from side to side. It touched the ground first with one wing tip, then with the other. Five or six times it repeated this sidewise swaying in slow motion. Apparently it was a threat or warning or intimidation performance. However the chipmunk paid little attention to it. Whenever, in its hurried zigzag progress, it headed directly toward the bird, the nuthatch darted in a quick retreat into the air. Its actions were new to us, a fresh discovery. But since that day we have seen this novel game of bluff repeated on a number of occasions, sometimes before other birds but most often in the presence of a chipmunk.

It was a chipmunk that was involved in another episode observed through the glass of one of our windowpanes. One day in early June I was leaving my photography darkroom in the basement when I heard a tapping and thumping at one of the small cellar windows. I assumed it was the brown thrasher I had seen fighting its reflection there a few days before. But when I peered through the pane, the creature that faced me on the other side of the glass was not a brown thrasher—it was a chipmunk. Like the bird, it was fighting a phantom rival. Continually it leaped at the windowpane, banged against the glass, clawed at its surface. When I was outside, later that morning, I saw the little animal, still in a rage, still battling its image in the mirror of the glass. I frightened it away. But as it ran off beside the foundation of the house, it came to three other basement windows. At each it paused and launched a fresh attack on the illusory rival that seemed keeping pace with its advance.

Where did the doughnut go? That was the great question remaining from one sequence of events that occurred just outside one of our living-room windows on a Sunday in February. The gusts of a winter gale pounded and shook the house. In a white smother the landscape was changing, the snow drifting everywhere. Seed I scattered in scoured patches was swept far downwind. Shining icicles, shaped by the gale as they were formed, ran along the eaves like a line of curved fangs. The weather bureau was announcing we were in the midst of one of the major storms of the century when our electricity went off. The furnace stopped. But blazing logs in the old fireplace kept us warm.

Outside, in the blizzard, most birds were lying low. But just beneath a living-room window, where gusts had swirled along the house and scoured the ground bare, we saw a bluejay. It was clinging with both feet to half a doughnut, one of several I had tossed from the back door. At every lull in the gale, it snatched mouthfuls of food. Each time a gust struck it, the bird tilted back, rocking with its feet still gripping the doughnut and lifting the far side from the ground. By using its stiff, out-stretched tail as a prop, it caught itself at the last moment and kept from toppling backward. Then, as we looked down, a tremendous gust howled around the corner of the house. The scene was obliterated in a whirl of blinding snow. Then the flakes settled; the air cleared. But the bluejay and the doughnut were gone.

Putting on a greatcoat and mittens and pulling down the ear flaps of my heavy winter cap, I went out to search for the doughnut. Although I hunted for more than 100 feet downwind, no trace of it could I find. There was no mark on the snow showing where it had fallen. Had it been carried away in the grip of the bluejay's anchored feet when the gust had hurled the bird into the air?

Once, years ago, *Life* magazine published a painting of a flying bluejay carrying an acorn held in one of its feet. At the time I shook my head. Bluejays transport food with their bills, not with their feet. I should, perhaps, have remembered Talleyrand's observation: "All sweeping generalizations are wrong, including this one." For, on this violent day of storm, under abnormal conditions, I may have encountered an exception to the rule. The fact that a smaller piece of doughnut, of lighter weight, in an even more exposed position, had not been moved by the gust seems to rule out the possibility that the heavier piece had been carried away by the wind alone.

During several successive summers, usually in July, Nellie and I have looked out the west kitchen window and observed a mystifying rite taking place in the dust of the lane. We first noticed it early one morning when we saw a woodchuck bending down, its red tongue rhythmically appearing and disappearing. Inexplicably it seemed to be licking the ground. For more than five minutes it remained in one place while an area about two inches square became dark with its saliva. After it had left I examined the spot. All the surface dust was gone and the harder ground beneath was scored by the animal's roughened tongue. Since then, always in June or July, we have seen not only other woodchucks but rabbits going through the same performance. Each time the animal left a little pit or depression in the lane where it had licked up the dust. On one occasion two small rabbits and an older one were all busily consuming dirt at the same time.

At first we assumed we were seeing something that, while it was new to us, was well known to others. But when I consulted Seton's multivolume *Lives of Game Animals*, I could find no mention of such activity. None of the other volumes in my library gave a clue to the mystery. And none of the mammologists to whom I talked at the American Museum of Natural History in New York had ever witnessed such a performance or heard of its occurrence.

Why were these animals consuming dirt? What impelled them to lick up the dust of the lane? My first assumption was that the ground there was impregnated with the salt we had sprinkled on the stone steps to melt off the ice after winter storms. But the animals always chose the far side of the lane in preference to the ground close to the steps, the ground that would be most likely to contain salt. And we saw them licking up dirt farther along the lane where none of the salt would be present. Moreover a friend of mine who lives in northern Maine tells me he has seen snowshoe hares eating dust along back roads in the forest, dirt roads where salt has never been applied. I wondered, for a time, if some animal smell was attracting the creatures. But twice when I lay down beside the place where they had left their dampened and darkened little pits, I could detect no unusual odor. All of the dirt eaters were vegetarians. The explanation for their action that now seems most likely is that the animals find in the dust they consume some needed mineral or chemical element, some beneficial addition to their ordinary diet.

Looking out the same window through which we had watched these dust eaters, we were fascinated, in the early summer of 1967, by another form of wild activity. For ten minutes or more we became absorbed in the strange behavior of a catbird.

When it caught our attention, it was balancing itself on the upper edge of the old carriage stone. This slab of gray rock, about nine feet long, five feet wide and six inches thick, tilts up at an angle just beyond the lane to the west of the house. Originally it formed the entrance to a carriage shed that has long since disappeared. Fragments of the foundation lie among the massed goldenrod of the weed jungle beyond the tilted stone. For our near-at-home wildlife, this bare rectangle of rock forms a center of activity at all seasons of the year. In winter it provides a natural feeding tray for birds. In summer rabbits and woodchucks sun themselves on it and chipmunks appear and disappear among the tunnels they have excavated around it. It is a kind of Stone of Peace, a common meeting place for many creatures, a place of amnesty in the presence of plenty. I remember one day after I had scattered cracked corn and birdseed on the stone I looked out and saw two chipmunks, six birds, a cottontail rabbit, and a woodchuck all feeding together.

But to get back to the catbird. Its wings drooped down and its tail was spread. It staggered along the edge of the stone above a drop of eighteen inches or so to the edge of the weed tangle. It appeared to be losing its balance continually, falling to one side and then to the other, arresting itself each time with an out-thrust wing. It teetered on the edge, tumbling forward, catching itself at the last instant. Minute followed minute while it continued its erratic course back and forth along the edge of the stone, looking downward all the time. The performance riveted our attention as long as it lasted.

At first sight, I wondered if the bird was anting. Then I thought it was putting on a broken-wing display. Its nest, containing young, was located less than thirty feet away in a lilac bush. Finally it occurred to me it might be in the presence of a snake. As soon as I emerged from the house, the catbird darted into the goldenrod. I moved silently up the carriage stone and peered over the edge. There on the ground below lay a coiled blacksnake. The parent bird, by its jerking wings and staggering movements, had been holding the attention of the serpent, distracting it from finding the nearby nest.

Since ancient times, stories of serpents charming birds have been common. I wonder if many of these tales were not derived from observing such a performance as I had witnessed. The activity of the bird catches the eye. It appears abnormal. The performer remains close to the snake instead of flying away. It seems held there, charmed, unable to leave. In reality, the reverse may be true—it is not the serpent charming the bird, it is the bird charming the serpent. By fascinating it, keeping it where it is, it immobilizes its search for the nestlings.

Edson Stocking once told me of seeing a similar distraction display put on by a robin in a peach tree. The bird's nest was on an upper branch. When a blacksnake began to climb toward it, the robin alighted on a limb between the serpent and the nest. It teetered and staggered, over and over almost, but not quite, losing its balance. Thus it diverted the attention of the blacksnake and led it away from the helpless young in the nest above. During a lifetime spent in the outdoors, Edson said, this was the only instance in which he had observed such a performance. It had been our good fortune to look from one of our windows at just the right time to observe this drama unfolding on the stage of the tilted carriage stone.

Three little woodchucks, born one year within a burrow hidden among the goldenrod, used the carriage stone as their playground all through the latter days of spring. There they romped, wrestled, pushed each other, fell off the stone and climbed back on again. Then, tired of play, they all would stretch out full length in the warm sunshine. One had a black face and two had white faces with black noses. The little animals were as friendly and playful as prairie dogs. In a world where there are a thousand woodchuck shooters to one woodchuck watcher, those weeks gave me a rare opportunity for observing their daily life. It was during these days that, from my study window, I witnessed something that I have never read about, something I observed on only this one occasion.

On the last day of June, about three o'clock in the afternoon, I saw the mother woodchuck sitting on the carriage stone with one of the young animals sitting beside her. It was now about half grown. Just as I was turning away, she reached down, grasped it by the fur of its rump, and lifted it off its feet. It hung head downward, partially curled. Previously when I had seen woodchucks transporting their young, they carried them, as gray squirrels do, gripped by the fur of their bellies and with the little animals curled up around their necks. But in those instances the carried groundhogs had been younger and smaller.

Supporting her heavier burden with head held high, the mother started off around the house and down the lane. At intervals she paused to rest. Each time, the young woodchuck uncurled itself and, reaching down with one of its forepaws, supported part of its weight by pressing against the ground. Then it would curl up again as the mother started on. For between forty and fifty yards, she continued down the lane. Then, turning aside, she pushed her way into a weed tangle beside the wall.

This tangle hid the entrance to the same empty woodchuck burrow that had been the goal of the cottontail rabbit when it dug through the snowdrift on that cold morning in December. A little later the mother reappeared. She came back up the lane alone. Several times in succeeding days I saw the young woodchuck near the hole to which it had been carried. It had been established in a burrow of its own. Similarly, during those days, I noticed one of the other members of the litter occupying a hole in the wild plum tangle at the eastern edge of the yard. Perhaps the original burrow had become too crowded as the young grew larger. At any rate the mother woodchuck had scattered the litter to various holes. Home ties were broken. The young animals were on their own. They all, apparently, remained where they had been taken. I never saw them on the carriage stone again.

In one of Gilbert White's letters, written from Selborne to Thomas Pennant, there is a sentence that has been described as "striking the keynote of the modern school of natural history." "Faunists," White wrote, ". . . are too apt to acquiesce in bare descriptions, and a few synonyms: the reason is plain; because all that may be done at home in a man's study, but the investigation of the life and conversation of animals is a concern of much more trouble and difficulty, and it may not be attained but by the active and inquisitive, and by those that reside much in the country."

Residing much in the country, we watch, at Trail Wood, the comings and goings of the creatures around us. We do not have to make field trips to encounter wildlife. We have only to look through our windows to see it. Through them, north and south, east and west, day and night, the year around, we can observe what the Selborne naturalist, in his quaint eighteenth-century phrase, termed "the life and conversation of animals."

CHAPTER TWELVE

Stone Fences

Across the river valley. Along the edge of our South Woods. Over the ridge beyond. This was the route pursued by a gaunt figure in old clothes and muddy shoes a century and a quarter ago. Plodding on foot back and forth across the length of Connecticut like some Ichabod Crane of science, following parallel lines about two miles apart, James Gates Percival was making his great pioneer study of the basic geology of the state.

Few stranger or more remarkable men have passed this way. The second son of a physician, he was born in Berlin, Connecticut, on September 15, 1795. All his strenuous labor, all his varied accomplishments were achieved in spite of ill health that dogged him throughout his life. So frequently did illness interrupt his studies at Yale that it took him five years to graduate. Yet he became a physician, a botanist, a geographer, a linguist, an editor, a geologist, and a poet of renown. At one time he was an assistant surgeon in the U.S. Army. At another time he was editor of *The Connecticut Herald*. At still another time he was professor of chemistry at the U.S. Military Academy at West Point. He surveyed iron mines in Nova Scotia and coal lands in New Brunswick. He assisted Noah Webster in compiling his dictionary. And, in the period before William Cullen Bryant, Percival was considered one of the leading poets of America. His works were published here and abroad.

With a high forehead, a prominent nose, a sallow complexion, and deep-set blue-gray eyes filled with brooding melancholy, Percival was virtually a recluse by the time he was thirty. In early manhood he fled from an engagement and he never married. Throughout his life he was unable to adapt himself to institutions or the irritations involved in working with other people. "I stood alone," he once wrote, "wrapped in suspicion and despair." His approach to practical matters was often characterized by timidity and indecision. When, in 1835, Percival accepted an appointment by the governor of Connecticut to conduct a geological survey of the state, James Russell Lowell made the mocking comment that the poet was now descending "from metrifaction to petrifaction."

What the governor and the legislature of Connecticut had in mind when Percival was appointed was a superficial survey concerned mainly with finding such new sources of wealth as mines and quarries. What Percival had in mind was something far different. It was anything but superficial. He set out with the grand design of examining all the ledges in the state and from these outcroppings determining the foundation rocks of the whole area.

His first idea was to make his survey on horseback. But he soon discovered a horse could not go where he wanted to go and that he was spending too much of his time caring for the animal. So through almost all of his patient, painstaking progress back and forth across the state, he traveled on foot. While staying at lonely farmhouses, after long days in the field, he often sat up until midnight recording his meticulous notes and carefully labeling his specimens. Soon after daybreak he was on his way again.

The time originally allotted for his survey passed and passed again while Percival continued his Herculean labors. The legislature grumbled. Where were the mines? Where were the quarries? The governor called him "a literary loafer." At the end of five laborious years the eleven manuscript volumes of his field notes totaled almost 1,500 pages. The rock specimens he had collected came from 8,000 localities in the state. Percival once calculated that in his travels on foot he had come in contact with every one of the more than 5,000 square miles that comprise the state. Finally, in 1842, his historic report, compressed to 500 pages, appeared in print. Entitled *Report on Geology of State of Connecticut*, it is a dry recital of factual data. But so detailed and accurate is this pioneer work that, more than a century later, the U.S. Geological Survey used it as the basis of a modern intensive study of the Hampton Quadrangle.

In his old clothes Percival was often mistaken for a vagabond and ordered off their property by landowners. Once, so the story goes, a farmer demanded to know what he was doing, examining rocks in his fields. Percival replied he was working for the state. The landowner responded that he was a taxpayer, he was helping pay his salary, and that he had a right to a complete explanation. Percival fished in his pocket, pulled out a nickel, and handed it to the farmer.

"That's your share of my salary," he said. "Now go away and don't bother me."

Later when he became the state geologist in Wisconsin—with better pay and with more appreciation—he was once supplied with a young man as an assistant to help him in the field. A few days later he burst into the governor's office declaring:

"I can't stand him!"

"Why?"

"He whistles and he throws stones at birds."

Probably the happiest time in Percival's troubled life came during his last years in Wisconsin. During the course of his geological explorations there he traveled more than 6,000 miles by buggy. In the remoter areas, he sometimes encountered parties of Chippewa and Winnebago Indians. His neighbors at Hazel Green, a village near the Mississippi in the southwestern part of the state, treated him with kindness and respect. His shabby clothes and odd behavior they accepted as signs of genius. It was at Hazel Green that Percival—gradually fading away, worn out by the long exertions his frail constitution had endured—died in 1856. His estate consisted almost entirely of books. They comprised, at that time, one of the largest private libraries in America.

In Connecticut a public subscription was started to buy this library and present it to Yale. But the effort failed and the memory of this remarkable many-sided man has grown dim with the years. At the New York Public Library, not long ago, I put in a call slip for *The Life and Letters of James Gates Percival*. The book I received was tattered and disintegrating. A note attached to it read: "This volume is printed on too poor paper to justify binding." It seemed an ironic comment on the life and ill-fortunes of the poet-geologist.

Along the country roads around Trail Wood—on the Old King's Highway, near Catden Swamp, on Hammond Hill along North Bigelow Road, on Griffin Road near the northern boundary of our land—I encounter the very ledges, perhaps a little more weathered now, that Percival examined with such care during his historic survey. In his report he noted that the rock of the long ridge running north to the west of us, the ridge behind which we see the sun sink at the end of each clear day, becomes increasingly ferruginous in the Hampton area.

In the immediate vicinity of our farm, a special feature of the rock formation is an outcropping of Hebron schist. This grayish crystalline metamorphic rock splits naturally into sheets and slabs of varying size and thickness. We find them scattered all across our fields. The most even and regular walls at Trail Wood and on the surrounding farms are formed of this rock. The schist is also prized for flagstones. In past times, when farmers near here heard loud thumping sounds along the road, they rushed out to drive off men who were stealing the flat capping stones of the walls and throwing them into trucks for sale in Providence or Hartford.

During several summer days one year, with a pedometer attached to my belt, I tramped along the walls that encircle our fields and border the road and wander away through the woods. On our 130 acres, I found, the combined length of these massive stone fences totals almost five miles. Each begins about two feet below the surface to provide a solid foundation below the frost line. Many rise as high as my chest and some as high as my shoulders. Leaning against these lichened walls in the summer sunshine, I often reflect on the back-breaking toil, the tremendous expenditure of energy, that went into their building long ago.

In pioneer times neighbors joined in; farmer helped farmer in wall building. There was always a plentiful supply of stones littering the rocky fields. And most of the early settlers of the region came from England, where stone walls and the construction of stone walls, in many areas, are part of the farmer's life. Toward the end of the main period of construction, which extended from about 1700 to 1875, gangs of men, expert wall builders, moved through the country erecting the stone barriers at so much a rod. At one time there were more than 100,000 miles of stone walls in New England. Year by year these elongated monuments to the labors of earlier men have been shrinking. Increased building activity and changes in agricultural methods have contributed to their destruction.

Green on gray, lichens decorate the stones of all our walls. Some cover the entire side of a rock as though with pale-green frost. Others, like the slow-growing *Parmelia* lichen, appear as rounded doilylike splotches. Some of these thin growths are older than the oldest trees around them. Walking beside our walls, a visiting lichen expert once examined one such primitive plant hardly larger than my outspread hand and calculated it was at least a century old. In Europe lichens more than ten centuries old have been reported.

In more than 16,000 different forms, these curious double plants—these cells of algae held together by strands of fungi; the fungi absorbing water and minerals, the algae manufacturing food for both—are constantly at work, unhurriedly disintegrating rocks to form soil. Carl Linnaeus observed two centuries ago that lichens form "the first foundation of vegetation." After foggy nights or in the wake of rains, a special beauty spreads along the walls as the lichens that splotch and plate the stones take on a deeper, richer, yet still delicate shade of green. During prolonged rainfalls some lichens can absorb water up to thirty-five times their body weight. On several occasions smaller lichen-clad stones from our walls have formed souvenirs given to visiting friends. One traveled as far away as Texas. In a shorter journey, another stone from beside the lane took its place among the rocks that form the cairn marking the site of Thoreau's hut at Walden Pond.

An entire book might be written about the natural history of an old stone wall. I remember a brown thrasher, after a night of rain, splashing about as it bathed in a small pool collected in a depression on one of the topmost stones beneath the hickory trees. Nellie once watched a long-tailed weasel wind in and out among the corridors of another wall. In crannies among the stones house wrens hunt for spiders. Along the flat tops, on late-September days, chipmunks, sleek and fat and ready for winter, scamper in sudden rushes, their cheek pouches crammed with food for stocking their burrows. Woodchucks sun themselves on the wall tops. *Polistes* wasps make their paper nests in spaces among the stones. Once, after a bitterly cold, windy night in early March, I came upon a dead bluejay pushed deep into a crevice. As death approached, it had sought protection from the wind and sanctuary from predators within the wall. During each winter many of the mice of the fields build their nests inside our stone fences. Here they are safe from foxes prowling in the night. Wherever there is a barway, we see the opposite wall ends of the gap laced together with a maze of mouse tracks in the snow.

For more than a week, late one September, a red squirrel took possession of the largest hickory tree beside the house. From morning until night it raced up and down the trunk, chasing away gray squirrels, harvesting the nuts, storing them deep in crannies in the wall below. Here they were beyond the reach of the larger gray animals. We watched it rattling up the trunk, darting out to some twig tip where a hickory nut hung, nipping off the nut, racing headfirst down the tree, disappearing into some crevice in the wall, popping out again. We saw it, on one occasion, bring down three nuts in almost as many minutes. The wall for twenty feet seemed crammed with its hoard of hundreds of nuts. It was, that winter, the wealthiest squirrel around.

On a June morning of another year, I watched a gray squirrel running along a wall where a number of the topmost stones were loose and tipped when the little animal leaped on them. Each produced a ringing sound that varied with its size and character. The squirrel seemed playing a rock xylophone as it advanced along the wall top. Henry Ward Beecher, in describing the ideal stone wall, concluded "it must be broad enough on the top to walk on." Every winter foxes and cats use our wall tops as roadways above the drifted snow. On occasion so have men. One of the stories descending from the blizzard of 1888 is how the Hampton stationmaster walked along the top of one of our boundary walls to get to work at the railroad on the ridge.

Although to Percival, during his survey of the bedrock of the state, they were of only passing interest, merely part of the "superficial" geology of the area, evidences of glacial action are everywhere in the Hampton area. Ours is Ice Age country. In its slow northward retreat, taking more than 4,000 years to cover the distance between Hampton and St. Johnsbury, Vermont, the melting ice deposited its detritus in many forms. Our hills and valleys were sculptured by the glaciers.

In 1966 the U.S. Geological Survey published a new *Geologic Map of the Hampton Quadrangle, Windham County, Connecticut*, prepared by H. Roberta Dixon and Fred Pessl, Jr. On this map two clusters of dots amid the winding contour lines show the location of glacial boulderfields on our land. One area where the great gray stones lie thickly scattered is just west of Juniper Hill; the other is on the ridge that runs north from the old chestnut stub at Old Cabin Hill. For the pioneers such areas represented the surface quarries where they obtained blocks of stone for the foundations of their buildings.

One huge boulder, with its top rounded like the back of a whale, contains a row of holes running in a straight line along its flank. They remain there from some long-ago abandoned effort to split from its blocks of usable size. Not far distant, another stone, a massive flat-topped rock, rests with one edge tilted up and held in position by a pile of smaller stones placed beneath it. It resembles a large deadfall. The mystery is explained by a habit of the early settlers. Before going to the trouble of cutting up and transporting such a rock, they carefully inspected the underside. With the weight of several men on the end of a slender tree trunk used as a lever, they pried up one side of such a rock and inserted supporting stones beneath it. At Trail Wood, although we have our boulderfields and our outcropping of schist, we have no limestone. This is a geological lack throughout Hampton. Given limestone, the list of our native ferns, without doubt, would have included several additional species.

That prominent feature of the Hampton landscape, the valley of Little River, came into being during past ages partly through the cutting of the stream and partly through the action of glaciers. Twenty thousand years have passed since the birth of the river; 10,000 since the last Ice Age. Glaciers, contrary to general belief, do not excavate valleys of their own. They deepen and reform valleys that are already in existence. The classic example of this is found in the famed Yosemite Valley of the Sierra Nevada range in California. It was started by the cutting of a stream but it was deepened as much as 1,500 feet by glaciers.

One outstanding component of the glacial landscaping of the Little River valley has already been mentioned. This is the kame, the rounded hill about two acres in extent, that forms the North Cemetery of the village. Like all kames, it was created by water pouring with its load of sand and gravel through a hole in the melting ice sheet. In 1819 the owner of this hill was Hesikiah Hammond. On November 9 of that year he deeded the area to the inhabitants of Hampton to provide "a decent and suitable burying yard to bury their friends." I once copied the provisions of the deed. They were explicit. The "inhabitants" are to maintain a "good sufficient lawful fence around said premises." They are not to "suffer any beast to enter and go at large on said premises, excepting sheep and calves." And no longer will the former owner be required to "subdue and keep subdued the bushes, briars and herbage growing on said premises." Taking note of its ancient use, the document points out that the hill "has been occupied for burying the dead beyond the memory of any living."

In the cool of the evening at the end of a summer day, there are few more pleasant and peaceful places than this cemetery hill that the glaciers made. It is set among wide fields and meadows. There are no human habitations close by. At the foot of the western slope of the kame the water of Little River spreads out into a beaver pond. Songbirds perch on the fences, alight in the fields, flit among the trees.

On several occasions, at such an hour of the day, the members of the Hampton Bird Club have wandered among the gravestones that record so many years of local history and have followed the paths that make their spiral ascent to the top of the hill with its view out over the surrounding valley. Once Nellie and I walked here with Roberta Dixon, the government geologist then engaged in her intensive study of the basic rock structure of the region. For so many summers did her work bring her back to Hampton that, in the end, she was considered one of the natives of the village. Everywhere, as we walked that evening, her educated eyes read stories in the tombstones around us.

The oldest were formed of local fieldstone. Then came a period when many of the markers were made of shining white Italian marble. Used sometimes to ballast ships in sailing days, this rock appears in tombstones of the time in many New England cemeteries. The identity of the material forming one large gray stone had special significance. It was Worcester phyllite, placed by geologists somewhere between shale and schist, a rock that is found in the vicinity of Worcester, Massachusetts. The stone marks the grave of Deacon Nathaniel Moseley, a prosperous Hampton farmer, the brother of the Reverend Samuel Moseley and father of Uriel Moseley, who married Sarah Hammond of "the House the Women Built."

It was the Deacon's custom, in the early days, to haul produce to Boston, using four oxen and taking a week to make the round trip. On these journeys south of Worcester he often passed a large stone highway marker. It occurred to him that here was a chance to obtain, without expense, an impressive monument for his grave. He hauled the gray slab home and when he died it took its place at the top of the hill among the other gravestones of the North Cemetery. For more than 150 years the Deacon's secret remained buried in the glacial drift of the kame. Then a woodchuck, one summer, digging its burrow at the base of the stone, weakened its hold. It tilted and finally fell flat on the ground. There it remains today.

By the light of the sunset, on that summer evening, we read the words cut into the phyllite: "In memory of Deacon Nathaniel Moseley who departed This life March ye 3rd 1788 in ye 73 year of his age. Memorio Mori. Blessed are ye dead that dieth in ye Lord. Their works do follow with a sure reward." When we reversed our gaze and looked at the other end of the stone, we encountered the words that had been hidden so long: "Right Hand Road to Boston. Left Hand Road to Worcester."

CHAPTER THIRTEEN

The Man in the Brushpile

I sit as though in a room with a hundred windows—small open windows of irregular shapes and varying sizes. Each is a chink or peephole in a circular wall of sticks around me. My seat is a chestnut plank. I lean my elbows on a benchlike table composed of two other planks of chestnut. Both the seat and the table are supported by a boxlike framework constructed of two-by-fours. It provides the central bracing around which are heaped dead branches gathered from the woods.

From a distance this gray mound of weathered sticks, about eight feet across and six and a half feet high, looks like a beaver lodge on land. During winter, when it is covered with snow, it resembles an igloo. As you draw nearer, you notice the narrow vertical opening of an entrance at the rear. Look inside and you see the seat and table enclosed within the hollow of a small circular room walled in and roofed over with interlacing sticks. Books, papers, maps, pads, and pencils lie scattered about. Working here in my brushpile study—a visitor once suggested I might call it my branch office—I frequently pause and gaze about me through the multitude of small openings that look out in all directions. I see without being seen. So, as I work, I keep track of the wild activity around me.

It occurs to me that, at this very moment, there are, no doubt, thousands of rabbits safely hidden within brushpiles and thousands of beavers snug in their lodges surrounded by sticks. But, in all probability, among all the millions of men inhabiting the globe, I alone am sitting within a hollow brushpile looking out. This has been an adventure in viewpoint that I have enjoyed during the summer days of a succession of years.

When I look through the small openings to my right, I see Hampton Brook flowing by twenty feet away. I see the lichen-decorated bark of tree trunks rising beside the stream. I see the gray rocks of the stone wall following the opposite bank and the green ascending slope of the pasture land beyond. If I turn and glance to the rear, I see the dense mounds of a blackberry tangle. If I peer through the peepholes to my left, my eyes encounter scattered bushes and trees. If I look ahead, banks of ferns, the waterfall, the pool below the falls, the wall that borders the woods and the swampy tract beyond it come into view. And looking up through the interstices in the interlacing sticks above me, I see the branches of the overarching trees. This is the setting, these are the surroundings, in which I spend my brushpile days.

One morning in the spring of 1962—when I was engaged in what seemed the happily unending task of going over Henry Thoreau's writings, choosing selections to include in *The Thoughts of Thoreau* —I became aware of a tearing, ripping sound outside the brushpile. I looked out to either side but observed nothing unusual. The sound was repeated. This time it came from directly overhead. I looked up. Projecting out above the brushpile extended the dead limb of a dying butternut tree. Bark was sloughing away with long dry ribbons of the cambium layer dangling down. As I watched, a catbird grasped one of these ribbons in its bill. It jerked and tugged until the strip broke free. Then it flew to the woods, a snuff-brown parchmentlike streamer fluttering behind it. Later, in the fall of that year, when the leaves had fallen, I came across a catbird's nest not far from the edge of the woods. Into it had been woven long brown ribbons of butternut bark.

Toward the end of an afternoon in August, during another year, I heard a loud continued splashing coming from the brook. It sounded as though a horse or a cow were fording the stream. But when I glanced in that direction I could see nothing to account for the sound. Laying down my pencil, I squeezed through the entrance opening to trace the splashing to its source. As I approached the edge of the brook, a flock of nearly a dozen robins rushed away. They all had been bathing together in the shallows at the foot of the bank nearest the brushpile.

Two robins, during one season when I was away from home, built a nest and raised a brood among the sticks of the brushpile. Another nest inhabited by a white-footed mouse—a ball of grass and vegetable fibers, about four inches in diameter, wedged among the crisscrossing sticks just beside my table—remained in place for months. In this elevated home, some three feet above the ground, the woodland mouse slept during the day, curled up in the soft interior, while I worked beside it. It never bothered me; but on a few occasions I could not resist the temptation to bother it. With the eraser end of my pencil, I gently poked the nest to see the bright eyes, the long whiskers, the large delicate ears of my brushpile companion as the little animal, awake in an instant, thrust out its head to assess the danger. On into the winter it lived there. When I went that way at dawn, I would find imprinted on the snow the maze of its tiny pawprints. Sometimes, following this record of its nocturnal activity, I would see where it had gathered seeds from dry weeds and how it had even carried its exploring as far away as the stone wall at the edge of the woods.

The top of my brushpile often formed a perching place for bluejays and chickadees. There they cracked sunflower seeds they had obtained from the feeders near the house. Not infrequently I had to brush away the empty hulls dropped on the table and seat before I could begin work. Gray squirrels also left the litter of their feasts scattered over the planks. When I was away, they jumped down from the branches overhead bringing nuts, and when they left, jagged fragments of the shells dotted my table. One hid a butternut in a crevice beside my seat. Another added the remains of a feast of apples to the nutshells.

I early learned to make deliberate movements when sitting within the brushpile in order to avoid alarming wild creatures outside. But even so I sometimes realized I had visitors only after they had taken fright. The turning of a white page, on one occasion, was followed by a sudden flutter of wings as some small bird that I had not known was perched above me rushed away in a panic. Another time I looked up when I heard scratching sounds overhead. At the movement a chipmunk that was surveying his surroundings from the summit of the mound of sticks went plunging down the side, chittering and squeaking in alarm. Then there was the July day when, after remaining unmoving minute after minute, engrossed in my work, I became aware of a slight rustling or scraping noise. Thinking it was outside the entrance, I swung around to look. A golden-brown form shot through the opening and plunged into the weeds. The woodchuck that lives in a burrow under a stump beside the brook had been exploring the interior of the brushpile. It had been under the very plank on which I sat when my movement sent it fleeing in terror. One thing was obvious—a groundhog depends on other senses than the sense of smell to warn it of the nearness of man.

At rare intervals some sharp-eyed bird would catch sight of me through a chink in the enclosing sticks. Once it was a rose-breasted grosbeak on a branch overhead. For some time I heard its sharp carrying single note, a squeak, loud and repeated. When I glanced up, I saw it twenty feet above me, hopping about, reversing itself on the branch, cocking its head far to one side, peering intently down, all the while repeating its call. A second instance of the kind occurred when a green heron alighted beside the waterfall. I had just laid down a book and it may have caught sight of that slight movement. It stretched out its long neck, tilted its head from side to side, seemed dissatisfied, flapped up to the limb of a hickory tree and craned its neck again. Leaves partially obscured its view so it flew to a dead branch on a red maple and there renewed its intense scrutiny of the mass of sticks and my now motionless form. All the time it was seeking to solve the mystery of a man inside a brushpile, it uttered a loud repeated clucking sound.

Those who have been much among wild creatures know that to them a human sneeze is almost as alarming as a gunshot. This was demonstrated one summer day when a flock of thirty-five or forty grackles alighted around my brushpile. For some time they hunted for food, carrying on a continual conversation in low grating sounds as they moved slowly across the ground. Then, after struggling in vain to prevent it, I gave an explosive sneeze. Instantly the brushpile was surrounded by a harsh clamor and an airy roaring of wings. The alarmed flock, rising as one bird, streamed in a headlong rush toward the protection of the woods.

The woods into which they fled, the woods I see west of the waterfall when I look north from my brushpile, was the setting of a dramatic period in the life of a beautiful animal some twenty years ago. The farm at that time was owned by Axel and Margaret Marcus. In one of the pastures at night, they discovered, some smaller animal was sleeping with their cows. Then they noticed odd things begin to disappear—Axel's rubbers, one of his work gloves, a pillow, a hoe. One morning they found where some unknown animal had burrowed under the chicken-wire fence to enter an enclosure holding laying hens. The poultry remained unharmed. But most of the mash that had been put out for the hens had disappeared. Weeks went by before they caught their first glimpse of the mysterious intruder. It was disappearing into the woods, a strikingly handsome collie dog, golden-tan with white feet, a white muzzle, white beneath, its tail a mixture of tan and white. Where had it come from? What was its history? No creature born in the forest seemed more wary.

Gradually, by putting out food where it would find it, they began to make friends with it. Axel climbed trees to watch where it went in the woods. Eventually he trailed it to a hidden den where, close to the waterfall, a maple had overturned in a windstorm and a cavern beneath its roots was screened by masses of overhanging ferns. Three small puppies that had been born in the woods occupied the nest.

And in a pile close by lay the rubbers, the pillow, the hoe, and the glove. Although the wild collie avoided people, she apparently found some comfort in things with a human scent just as she found comfort in sleeping among the cows. Obviously it was on some farm that she had spent her earlier days.

The puppies were brought home and little by little Gypsy—as the Marcuses had named the dog—lost her fear of humans. But it was noticed she was always terrified by thunder and lightning. She was made uneasy even by branches waving in the wind. Then, one day, a man from Abington, five miles away, called at the farm. When he saw Gypsy, he exclaimed:

"Why, that's my dog—the one that disappeared in the storm!"

More than a year before the tan and white collie had been sitting beside him as he stood in an open barn door during a violent thunderstorm. In a blinding flash, a bolt shattered a tree nearby. When he looked around, the dog had vanished. For days he hunted and inquired without finding any trace of her. The lightning-terrified animal, a pedigreed collie, had run for miles until she was completely lost and then had lived, as best she could, in the woods. Seeing she was contented and well cared for, the former owner left her at Trail Wood.

During the first winter, although the Marcuses made a snug nest for her in the middle shed, leaving the door open as we do for the phoebes in summer, Gypsy slept outdoors, curled up in the snow. As time passed, however, her wild ways decreased. She became more confiding and tame. She regularly slept on a blanket on a bedroom floor and she and the family cats fed amicably from the same plate. For another decade this "lightning dog" continued to live on the farm. When she died—not long before we came to Hampton—she was buried near the foot of the slope in Firefly Meadow. One of the three puppies born in the woods was still alive at Trail Wood when we arrived.

On late-summer days, from the loopholes that surround me in my brushpile, I see the red of cardinal flowers, the yellow of jewelweed, the white of boneset and mountain mint. In the early spring violets bloom outside the mound of sticks and even around my feet within its shaded interior. As the years have passed, a tinging of green has spread across the planks of the seat and table where tiny mosses have gained a foothold in cracks and damp depressions in the wood. Blackberry canes have pushed their way upward among the sticks while fungi and lichens have increased on the oldest branches. Each addition of the kind has added to my camouflage.

One curious thing I have noticed in connection with my brushpile. Whenever I enter it, I leave behind the biting insects, the deerflies and horseflies, that fall upon me again when I emerge into the open. Nor can I remember ever being bitten by a mosquito while sitting inside. But various other insects, from time to time, keep me company. Diminutive mayflies occasionally drift in through the openings and shed their skins while anchored to the framework inside the brushpile. One left its delicately translucent brownish husk attached to my table. It reproduced every feature of the frail insect, even to its slender threadlike tails. So light was this chitin shell, this mold of an insect, that when, in leaning close to examine it minutely, I breathed upon it, it rolled across the plank like a Lilliputian tumbleweed.

The sounds of the seasons, from spring to fall, are woven into my memories of the hours spent within this hideaway. There was the humming of bees among the mountain mint, the liquid monologue of the waterfall, the steely chorus of the meadow insects, the rushing of wind among the summer leaves. In the hot still days of August, the shrill cicada's song would soar above me among the treetops to end in a descending buzz and a dying sizzle.

Toward fall the gray squirrels added another sound. As they worked among the branches of the hickory trees, I would catch the ripping noise of shucks discarded from the nuts descending through the foliage. Later, with my ears, I would follow the progress of chipmunks gathering seeds and nuts among dry fallen leaves. A chipmunk running among dry leaves sounds almost as big as a deer. On one October afternoon, when I looked out, I saw one of these little animals racing down a slanting mossy rock at the water's edge. It was in pursuit of a rolling hickory nut. Just before the prize plunged over the edge into the pool below the waterfall, the chipmunk retrieved it.

Sometimes, far away, I heard the crowing of a rooster. It seemed as wild as the voice of the wildest bird. And always, sometime during the day, I would catch the calling of the crows. Anyone who fancies that crows utter only a "caw" should have been in my brushpile on those June days when a family of these birds alighted among the brookside trees. Sitting silent and unseen, at such times, I became absorbed in noting the variations in volume, in inflection, in quality, in tone, in tempo, in emphasis—in a word in how wide a vocabulary they employed. I would listen to sounds that were harsh and guttural and sounds that were soft and caressing, the intimate family talk of the birds. They gave the impression of engaging in argument, dialogue, repartee. Oftentimes crows seem to converse where other birds call or sing. It is understandable how people have imagined that crows hold trials, condemn members of the flock, debate what course of action to take. Once when two crows attacked a third close by, the harried bird uttered a hollow rattling woodpeckerlike sound such as I have rarely heard before or since.

A whole chapter, I suppose, might be devoted to the birds and the bird sounds around my brushpile study. For more than two weeks, in late June and early July, one year, I followed the fortunes of a family of ruffed grouse that spent its days around my mound of sticks. I first encountered the little group on an afternoon of gusty wind. Weeds, bushes, trees writhed and lashed in the blasts. The rush of the gale through the foliage drowned out my footsteps as I came along the path beside the brook. Just before I reached the brushpile, an adult grouse shot up out of the weeds. It appeared to become entangled in the vegetation, pitched down, floundered ahead. I was led by its faltering progress almost to the edge of the woods. There it lifted cleanly into flight and disappeared among the trees. As I turned back six baby grouse, hardly larger than starlings, rose on short laboring little wings from around the brushpile. Their squealing cries cut through the booming of the wind in the treetops. Fluttering just above the weeds, they scattered in all directions.

During the days that followed I often heard the grouse feeding, unseen, close by. I watched them trailing out of the woods. I saw them sunning themselves in an open space at the foot of the wall. Sometimes both adults accompanied the little band, the older birds uttering an almost continuous series of small soft calls. As the baby grouse grew larger and their wings grew stronger, their flights became more extended. On one of the last occasions on which I saw them, one of the brood burst up out of a mass of jewelweed near the waterfall. On straining wings it climbed higher and higher, above the wall and into the woods, to land on a dead branch twenty feet above the ground in a red maple tree. There it perched for several minutes. It peered about, lengthening its neck until it suggested a green heron in miniature. Then trusting itself once more to its wings, it launched out on a long downward slant that carried it farther away to the woodland floor.

The explosive splash of a kingfisher plunging into the brook for dace is a sound that I hear, from time to time, in the spring. How these birds avoid dashing out their brains on the rocks of the shallow pools is an unending marvel to me. At this time of year, when they are feeding their nestlings, they haunt our brook in search of the smaller fish. Always I see them fly away with the dace held head outward in their bills—in the right position for stuffing them down the throats of the young. Along the brook and past my brushpile, singing as it flies, the Louisiana waterthrush wings by to disappear into the woods where it nests beside the stream. Through my field glasses, one morning in spring, I saw one of these warblers, in searching for food, busily turning over wet soggy leaves beside the dam.

Probably the most unusual bird activity observed around this waterfall occurred about one o'clock in the afternoon on April 5, 1964. A pair of wood ducks came sweeping in to alight in the pool above the falls. For a time they floated or swam about along the still edges of the water. Then the female paddled out into the main current, approached the waterfall, hesitated a moment on the brink and then tobogganed down the cascading water into the smaller pool below. A minute or two later the male appeared at the top of the foaming water and made the same thrilling descent. So close to my mound of sticks, the two waterfowl seemed playing games, enjoying a carefree interlude.

Each autumn, when the pool below the falls is transformed into a mosaic of floating colors fallen from the trees, my pile of dead branches becomes spangled with the brilliant reds and yellows of the descending leaves. Later, when the limbs above are bare and winter cold has settled down and I have abandoned the brushpile for the year and snow has transformed it into a mound of shining white, I come that way to observe the tracks wild creatures have left around it in the snow.

Squirrels, dropping down from tree limbs, head away in all directions, leaving trails radiating outward from the brushpile like spokes from the hub of a wheel. One morning I examined the tracks of a large deer that had circled the mound in the night. I saw where it had stopped and changed its position to look inside. And I observed where something had alarmed it and it had headed for the woods in great bounds, twelve or fifteen feet long, the hoof marks bunched wherever it landed. Foxes, raccoons, mink on other occasions have paused to inspect the snow-mantled brushpile. When I followed the catlike prints of one fox, I saw where it had taken to the open road of the ice-covered brook and then had surmounted, in a single leap, the mass of the frozen waterfall.

Twice in its life, my pile of dry branches survived the tail ends of hurricanes. The raging winds that stripped away the leaves and plastered them against the sticks, whistled through the innumerable openings in the brushpile and left it unharmed. Once, nearby, the whole top of a decaying maple snapped off in a storm. As it lay on the ground, I walked among its upper branches, the very ones where, on so many summer days, I had watched crows and robins and flycatchers perching in the sunshine. Over the years, other limbs crashed down around my mound of sticks without hitting it. It lived a charmed life.

At the close of the year 1972, when the leaves fell and snow covered the brushpile, we had no inkling of disaster. But while the seasons had passed, while I sat within my many-windowed room looking out, the old butternut standing beside it, the same tree that had supplied the catbird with its nesting material, had grown older. It had lost its last leaves, then its bark. Its hidden roots had slowly decayed. Late in March, in 1973, the last of the winter gales, howling out of the northeast, raked across our fields. Its long gusts shook the house. On the afternoon of the thirtieth of that month Nellie and I walked to the waterfall to see a clump of pussywillows, golden with pollen, surrounded by the humming of early honeybees. It was thus that we came upon the disaster left by the storm. The butternut lay prostrate. It had fallen at last. And as it had fallen it had crashed squarely across my brushpile. Sticks, table, seat, framework all lay in splintered fragments, demolished beyond repair. Like a shipwreck or a house leveled by an earthquake, the disaster was complete.

So those ever-memorable hours spent in the heart of a hollow brushpile came to their termination. That gray mound of sticks had provided for me a quiet once-in-a-lifetime experience, an adventure unique. It had lived its life and come to its end. I never replaced it.

CHAPTER FOURTEEN

The Old Woods Road

The first thing you notice as you cross the Starfield and near the edge of the North Woods is how sounds echo from a small bay in the wall of the trees. Your voice is caught up and flung back, seemingly magnified. It is like singing in a shower bath. Sometimes, on still mornings in summer, when the fire whistle blows in the village to our south, the sound appears to come from the north. The echo from the woods is clearer than the wailing note that reaches our ears from its source. Each year, as autumn draws on, the echoes lose their strength, then die entirely as the foliage falls.

Almost at the center of this bay of echoing trees, a small gate opens on a narrow woodland path. Resilient with moss, noiseless, dustless, it winds under the shade of swamp maples, crosses Fern Brook on three gray stepping stones, circles the foot of Twig Hill and follows the Old Colonial Road as far as Hyla Pond.

Here, below the ridge where the silvered stub of the great chestnut rises, the path swings to the right, turning north on a meandering course through half a mile of the wilder, more remote reaches of our land. It advances along the traces of an old farm road, an old woods road, long abandoned. Among all the trails of Trail Wood it is our favorite. It is the one we follow most often. Everywhere along its length there are little detours, old friends among the trees, special plants we visit at different seasons of the year. From it we look into deeply shaded little glens that always bring to mind words of an unknown song I heard but once and have remembered all through the decades since:

"Somber woods, ye glades dark and lonely."

For the first hundred yards or so, the remnants of this former road follow the base of the ridge. Here it is always coolest in summer, coldest in winter. Here the snow remains unmelted longest in the spring. We sometimes refer to this section of the trail as the Refrigerator. Above the path, the whole slope of the ridge is carpeted with masses of Christmas ferns whose fronds retain, all through the winter, their dark and glossy green.

The acres drained by Fern Brook, acres that border the beginning of the Old Woods Road, well substantiate the appropriateness of the name of this little stream. The area is rich in mosses and ferns. Growing here we find twenty-one of the twenty-six ferns native to Trail Wood, including our only stand of broad-beech ferns and our single plant of that rare variety, Boott's fern.

A little farther along the trail, where a small valley extends in a curving sweep between two wooded hills, we come upon the red-stemmed form of the lady fern. And when we turn aside to follow this valley, it leads us to a slope where we find rooted the only fragile fern we have at Trail Wood. Its little fiddleheads are among the earliest we see in spring. With its brittle, easily broken stem, it seems ill-fitted to survive. Yet it is one of the most widely distributed of all the ferns. It is found from Greenland to New Zealand and even high in the Andes. Its hardiness is its strength.

Friends, more knowledgeable than we, have identified more than fifty mosses along our woodland trails. Other species, no less beautiful, remain anonymous. At one moist and shady spot where a stump has moldered away, a patch of moss, emerald green, forms a mass so soft and dense it suggests a velvet cushion discarded beside the trail. This is the whip-fork moss, *Dicranum flagellare*. Examine it in early summer and you find it covered with upright gemmae, budlike bodies of reproduction complete with stems and leaves. Run a forefinger across the cushion and you observe how the small gemmae break off at the slightest touch. Each of these thousands of tiny reproductive bodies is capable of producing a whole new stand of moss.

Close to the start of the Old Woods Road, one hot afternoon in July, Nellie and I noticed a thin column of running ants. As far as we could see—over the moss and around the ferns—they were keeping to a definite line of march. Some were going one way, others were hurrying in the opposite direction. Most were small and dark. But scattered among them were reddish ants, two-toned, with abdomens of darker hue.

What we had encountered on that summer afternoon was an insect slave raid. The ants and their slaves from one colony were attacking the ants of another species and bearing away their pupae. From these would come new slaves that would live and work in the nest of their captors. Usually such raids commence in the morning and end in a kind of triumphal homeward procession in the afternoon. Almost all the reddish ants we saw were carrying in their jaws small whitish objects that looked like grains of rice. These were the living pupae, the booty of the war party. The reddish ants, slaves themselves, were transporting these future slaves back to the underground city of the conquering insects.

I remember we turned aside to follow their line of march. It crossed the road, wound in and out along the swamp edge, advanced over dead leaves, around mossy rocks, under spicebushes. In one place it crossed over a small gray feather that had fallen from some molting bird in the woods. Then, almost on a compass course, it climbed the steep ascent of the ridge to end at a large ant nest near the summit. I paced off the distance from this nest to the spot where we had first encountered the raiders. The route these small creatures had followed, I found, extended for about a tenth of a mile. In their raid, coming and going, the insects of this invading army had covered nearly a fifth of a mile. During a subsequent year, in this same area, again in the afternoon but this time in August, we encountered the ants from this same nest making a similar raid almost as far from home.

In our region, the woods normally are bare six months of the year and clothed with foliage the other six. During summer days we advance with the path of the old road mottled with sunshine and shadow. No two of the streakings and splotchings of light descending through openings among the leaves appear identical in size or shape. Rarely are they of the same brightness. In one spot the light may be dulled by a dark carpet of moss, in another it may be brightened by the shine of tan fallen leaves remaining from the previous year. Our minds, without being completely conscious of the source, are pleased and diverted by this endless variety.

In and out of the sunshine drift the colors of butterflies. Chipmunks scurry over the leaf carpet of the woodland floor. Ovenbirds set the open glades to ringing. In his *Malay Archipelago* Alfred Russel Wallace notes that he discovered the best places for collecting natural history specimens were along primitive roads and trails in the jungle. We have repeatedly noticed that in the denser portions of our woods we see less wildlife activity, less of nature in action, than when we follow such an open path as this old woodland road.

On such a walk, our eyes, roaming to either side, expand the area we cover. We advance, usually, in a kind of two-step, taking a couple of steps, then pausing to observe something that has caught our attention. By following all the turnings of the same familiar path, we see the same scenes at different hours of the day and in all seasons of the year.

We see them in the spring, when dark little butterflies with the charming name of Dreamy Duskywings whirl in the sunshine under trees clothed in their earliest green. We see them in the hot still days of midsummer, when we hear the pewee and the shrill cicada's call and when along the trail we come upon feathers of various forms and colors shed by molting birds. We see them in autumn, when we scuff through fallen leaves and find small oak apples, crimson insect galls, that have dropped from trees and lie scattered on our path like ripe red cranberries. I once counted 119 in the space of seven paces. As much as 50 per cent of these red spheres may be composed of tannin, that substance that makes acorns bitter and unripe persimmons puckery to the tongue. We know the Old Woods Road in winter, when we sometimes see where foxes have trotted along the same path we are using and where, on occasion, we follow some line of blood spots on the snow until it ends at a tree where an owl in the night has alighted with its victim.

The combined memories of our walks telescope together—as in a lapse-time moving picture—the shifts, the continual changes that occur in the scenes around us. This woodland that appears so solid, so set, is in fact almost fluid in its endless alteration. An old tree falls; a new sapling springs up. I recall that once as we went up the trail we paused to look at a bird perching in a long-dead tree. Fifteen minutes later, when we returned, the trunk lay prostrate, the top shattered across our path. It had fallen, had been subtracted from the standing trees of the woodland, in that short lapse of time.

After every windstorm we find the old road strewn with dead branches. So the trees die in fragments. Occasionally, in falling, a tree becomes hung up in the limbs of another tree and remains tilted steeply across the path. To our neighbors it is known as a "widow-maker." We always pass by such a tree warily, especially on windy days.

Above a deep and extensive valley filled with ferns and mosses, springs and rivulets, skunk cabbage and false hellebore and the soggy expanse of a red maple swamp, the road curves around a hill clothed with the scraggly bushes of witch hazel. Always along this part of the trail, in the latter days of autumn, we stop to examine the last flowers of the year, the loose clusters of tiny yellow ribbonlike petals that appear along the branches and at the twig ends of the bushes all across this Witch Hazel Hill.

From them, a year later, mature the shining black seeds. During our first September at Trail Wood I brought home from the North Woods a branch of witch hazel bearing several clusters of seed capsules. Late that night I was awakened by sharp rapping or striking sounds repeated again and again. The next morning I found that the capsules had opened and pressure from within had shot the seeds out as an orange pip is propelled when squeezed between a thumb and forefinger. They lay scattered over the floor of my study where they had fallen after striking the walls. On occasion such seeds are hurled through the air as far as forty feet.

Just such an experience as mine is recorded by Henry Thoreau in the twelfth volume of his *Journal*: "Heard in the night a snapping sound and the fall of some small body on the floor from time to time. In the morning I found it was produced by the witch hazel nuts on my desk springing open and casting their seeds quite across the chamber, hard and stony as these nuts are." The date of his entry was September 21, 1859. Thus 100 years later, in the same month of autumn, history had repeated itself at Trail Wood.

On a February day when the wind blew in cold sustained gusts from the north, we rounded Witch Hazel Hill and heard a shrill squealing sound among the trees ahead. It cut through the confused roaring of the wind. Then it was repeated in a different key. That was the beginning of a strange two-year arboreal contest that we watched with constant interest.

When I traced the sound to its source, I found two trees, a red maple and a pignut hickory, standing about ten feet apart. During a gale in the night a dead lower limb on the maple had broken partly off. Its outer end, about two and a half inches thick, had lodged in the V where a hickory limb joined the trunk. Each gust that rocked the trees sawed the dead branch back and forth across its support. Already it had worn a noticeable groove in the hickory wood and a fine white line of powdery material clung to the bark below. At each movement the ear-piercing complaint of the rubbing fibers filled the woods. How long, we wondered, would it take the dead branch to saw its way through the living limb? The answer to that question, we knew, depended on the number of calm and windy days.

All the rest of that winter, all the summer and autumn and winter that followed, all during the seasonal sequences of a second year, we watched the progress of this odd duel beside our trail. Each time we came around the hill and stopped beside the trees, we expected to see it ended. But the wood of the maple branch, where it made contact, was soon worn to a polished smoothness. It lost much of its abrasive quality. And the living hickory limb, even though it was only about an inch and a half thick, was as tough as rawhide.

On days when a succession of gusts struck the trees, we would see the dead branch moving back and forth like a bow sweeping over the strings of a violin. At each stroke a sustained wail or screech would rise from the scraping wood. Even in light breezes the shift of the limb produced small squeaks or high-pitched squeals, abrupt and short-lived. Under certain circumstances, the complaint of the maple rubbing on the hickory came to our ears as a low buzzing sound like the drone of bees.

Slowly, steadily, as we watched the duel continue month after month and season after season, the groove in the smaller branch grew deeper. Almost imperceptibly its strength decreased. Two years and an additional month went by after our initial discovery before the end came. It came suddenly. But not in the way we expected.

Soft wet snow fell early on the morning of that late day in March. When, in the afternoon, we plowed up the trail to the spot we found this added weight had been too much for the hickory limb. It had bent downward at the groove, allowing the branch it supported to slide off and drop to the ground. So the long struggle ended. We had seen it at its beginning, on February 22, 1967, and we had seen it at its end, on March 25, 1969. The hickory limb had survived. It remains permanently bent downward, the groove clearly visible at its base. We look up at it each time we go by.

A broken branch of another kind produced a different memory of this portion of the trail. Sauntering along in the morning sunshine, on a day in April, we suddenly became aware that the air around us was filled with the scent of wintergreen. We paused and looked about us. Close to our path, sap oozed from a dangling limb on a small black birch. It is from such trees that the wintergreen flavoring that is sold commercially is distilled. How many other scents are linked in our minds with walks along the Old Woods Road—the balsamlike smell of hickory buds in the spring; the pungent perfume of crushed spice-bush leaves; and all those rank stirring animal smells that rise at various places along the path, smells that would mean so much more to fox and weasel and hound than they mean to us!

Our surroundings grow more wild and lonely as we advance beyond Witch Hazel Hill. The signs of deer increase. We see hoofprints in the path and trails leading away among the trees and through gaps in old stone walls. We notice saplings where bark is gone and the wood is polished by bucks rubbing velvet from their antlers. And during autumn days we come upon small areas of bare ground grooved by the pawing forefeet of the males. These are their "scrapes," warnings to rivals.

After we pass the end of an ancient wall, we dip among oaks and maples on a long descent to Hampton Brook. Originating in the Natchaug State Forest on the other side of the abandoned railroad, this stream descends for about two miles to its junction with Little River. Fully half of its winding length is contained within our acres. Early users of the road forded the brook across the shallows just above a cascade where the water tumbles down among moss-covered rocks. Set beside this cascade, among violets and where the small rounded leaves of the partridge berry carpet the ground, we find a chestnut plank supported by flat stones stacked like pancakes beneath either end. On this crude bench, during summer noons, we often sit, with the stream foaming over the rocks before us, and eat a picnic lunch. Although the Old Woods Road continues on for another quarter of a mile, winding in a slow ascent to Griffin Road, this spot, the Brook Crossing, is the center of interest for us in this northern part of our land.

Everywhere along the paths of Trail Wood wildness seems near at hand. But nowhere else do we feel so remote from the world as here beside this woodland brook as it traces its serpentine course among the mosses and ferns and trees. We might be in the midst of a scene in the Adirondacks or far back in the woods of northern Maine. As we remain silent and motionless, we expect, at any moment, to see a mink come hurrying on a twisting track along the brookside or a deer peering at us from among the trees. So wild does this setting seem that one August day I even brought along an aluminum pie tin and at the little gravel bar above the ford panned for gold. I was not disappointed. That is, I expected to find no gold and that expectation was realized.

Resting immobile on our bench beside the stream, we frequently listen to those two earliest musical sounds on earth—the liquid rushing of the water and the airy sighing of the wind. In winter we hear the creakings, the tappings, the shrill complaint of the trees. In summer the sounds are, in the main, softer, more melodious—the rustling of the leaves, the singing of the birds. Above us we often hear the little white-breasted nuthatch hitching along tree limbs, talking to itself. We catch the "Chipchurr" of the scarlet tanager. And always, somewhere in the hot woods of midday, the voice of the red-eyed vireo goes on and on. But I suspect the most characteristic living voice at the Brook Crossing is the chipping of the chipmunks.

With ample food and old stone walls, these are the chipmunk woods. Some summers the lively striped rodents are everywhere. We hear their squeaking cries of alarm. We see them scurrying over the leaves. We find remnants of their meals scattered across the plank of our bench beside the brook. And we listen, especially as summer draws to a close, to the endless chipping repetition of their territorial calls. It was here, as I mentioned in an earlier chapter, that we heard one chipmunk continue, without a pause, until it had repeated its sharp little call 536 times.

For several years after we came to Trail Wood, we were puzzled, in the earliest days of autumn, by a low, hollow "Cuk! Cuk!" that echoed through the woods. It suggested the repeated call of a black-billed cuckoo. But it had a quality of its own. Once I flushed a grouse near the spot where the sound seemed to originate. I wondered for a time if it might be a seasonal call of these game birds. Then one October morning, along the Old Woods Road, I made a sudden rush toward the sound. A chipmunk I had not seen before raced away up the slope with its ringing chitter of alarm. Higher up it stopped, mounted a low stump, and its "Cuk! Cuk!" call began again. The mystery had been solved.

Some years ago an ornithologist studied the feeding habits of warblers. He made the discovery that one species will habitually begin its hunt for food at the trunk of a tree and work outward to the tip of the limb while another species will begin at the end of the branch and work inward. Thus the search is made from two directions. What one misses the other is likely to find.

This wisdom of the warblers also applies to walking a woodland trail. To see the most, go both ways. So when we turn back along the Old Woods Road on our return journey we always wonder what new object will catch our eye, what fresh adventure, small or large, will await us. What Nellie saw, one autumn afternoon, and the events that resulted from her discovery, will be told in the following chapter.

CHAPTER FIFTEEN

Beavers in the Moonlight

Among the witch hazels above the swampy valley on the Old Woods Road, that early October afternoon, Nellie stopped and listened. In the trees below, she thought she heard the calling of rusty blackbirds. In descending to investigate, she made an exciting discovery. The whole lowland, perhaps three acres in extent, was flooded. Where there had been a swamp in the spring, there was a pond in the fall. And it was no ordinary pond. It was a pond produced by the earliest dam-builders in America. How much wilder our wild acres seemed: beavers had come to Trail Wood!

At the end of its wandering course through the lowland that had become a pond, Hampton Brook flows on through a gap in a boundary wall. It was here that the beavers had blocked its flow. Using the solid wall on either side of the gap as a foundation, they had erected their barrier. Then they had filled in the gap with mud and sticks. When we measured the dam, we found it was 125 feet long with nearly 100 feet of its length braced by the heavy stonework of the wall. Already, in preparation for their first winter, the colony had begun the construction of a lodge. Out in the pond, the mound was rising like a medieval castle surrounded by a wide protecting moat.

Since that day of Nellie's discovery, in all the months of the year, we have walked along this brown water in the heart of our woods. We have rested on beaver-felled trees when the mercury stood above ninety degrees and we have plodded through snow beside the ice-covered pond when the thermometer registered zero. The age-old activity that fascinated us was familiar to the Indians and the French trappers and the first settlers in America, to all those European wanderers who, in the earliest exploration of interior North America, were lured farther and farther into the wilderness by their search for beaver pelts. For our pond-dwellers continued on into this atomic and satellite age with ways unchanged.

Weeks went by before we first sighted a beaver. But more than once we heard the loud warning slap of a paddle-shaped tail striking the surface of the pond. It was followed, as the unseen animal dived, by a splash like a boulder plunging into the water. We had seen no beaver. But a beaver had seen us.

It was November—the month of the Beaver Moon—before, in rather unusual circumstances, I caught my initial glimpse of one of these wary creatures. Toward the middle of that month a freeze covered the pond with half an inch of ice. About four o'clock in the afternoon, on the seventeenth, I had finished nailing up signs on trees, warning that the beavers were protected. I was standing not far from the edge of the pond opposite the lodge. Like a rifle shot, a sudden explosive crack came from that direction. A moment of silence followed. Then a second crack. Out on the pond, no more than ten paces from where I stood, the ice was heaving up and down. Long white lines of fracture radiated away. Then the ice burst upward. Fragments shot into the air and skidded over the frozen pond. Bubbles cascaded up through the water and the dark dripping head and back of a beaver appeared in the opening. Standing in a shallow place in the pond, it had heaved rhythmically upward with its back until the ice gave way.

I froze motionless behind the maze of small twigs on a high-bush blueberry. For a time the beaver rested. It gazed about it without seeing me. Then it appeared to catch my scent. Four or five times, it lifted its broad nose and sniffed the air. Finally, unhurriedly, without alarm, after looking around it for a last time, it lowered its head below the ice and pushed with its strong webbed hind feet. I could see it for some distance swimming under the ice toward the lodge.

A powerful, but not a particularly fast swimmer, a beaver's average speed through the water is about two miles an hour, approximately that of a swimming dog. However its ability to remain submerged is outstanding among fur-bearers. Beavers have been known to swim half a mile underwater and to stay below the surface for as long as fifteen minutes. When a beaver dives, valves in its nose and ears close automatically. Its unusually large lungs and liver provide it with an exceptional supply of air and oxygenated blood. In winter it can renew its oxygen supply by expelling large bubbles of air and then breathing them in again after they have been purified by contact with the water and the underside of the ice.

The large mound of the lodge toward which my beaver swam provides what is essentially a burrow inside a pile of mud and sticks and vegetation. The circular space of the central room, about two feet high and six feet across, is clawed out and the projecting sticks are gnawed off until the right size has been attained. Even in zero weather, the heat given off by the animals is sufficient to warm the interior. The entrances to a beaver lodge, usually two, sometimes in large houses as many as five, rarely one, are always underwater. One of the folklore tales of the pioneer West is how John Colter, famed for his adventures in the Yellowstone country, escaped from pursuing Blackfeet Indians, in 1809, by plunging into a beaver pond and coming up inside one of the largest of the lodges.

As winter drew on, while ice hid the activity of the Trail Wood beavers from our eyes, we could picture them enjoying ample leisure within the safety of the lodge after "working like beavers" in the fall, storing up food in the form of an underwater cache of poplar poles. When hungry, they could swim to the submerged pile of sticks, cut off a desired piece and swim back to the lodge again. In the course of a full meal, a winter beaver consumes about three pounds of aspen bark. How, we wondered, do the animals gnaw through a submerged pole without having water rush into their mouths? The answer, I found, lies in the looseness of a beaver's lips. They can be drawn tightly together behind its four protruding cutting teeth. Thus its throat and mouth are sealed off while it is gnawing underwater.

Along the edge of the pond, after ice had formed but before the ground was covered with snow, we wandered among the stumps of aspens that now lay hidden in the submerged pile of the winter food supply. Most were less than six inches in diameter. In places we would come upon half a dozen little heaps of chips, almost equally spaced, where the trunk of a felled aspen had been cut up into manageable lengths for towing to the sunken food pile. A maze of paths led down the hillsides to the pond. Along them the beavers had dragged their poles, flattening down the weeds. At one place, we found a path that had been worn for nearly 100 feet up a slope to the edge of an open meadow. There a clump of aspens had been felled. Their trunks, after being cut up, had been dragged down the path to the water.

In this activity the all-important tools of the animals are their four bright-orange cutting teeth. They are self-sharpening and grow rapidly. The upper two provide anchorage and leverage while the lower two, in a powerful upward bite, chisel out the chips. Working in this way, a beaver can cut down a tree with a five-inch diameter in about three minutes.

In our new colony, the food was exclusively bark. In some old established ponds, the animals sometimes consume other foods as well—the rootstocks and buds of waterlilies, raspberry canes, mushrooms, duckweed and berries. There are even records of "grass beavers." Living in localities in the West devoid of trees, they feed chiefly on grass. But always the beaver is a vegetarian. It never takes fish or frogs or other animal fare.

On the second of February, that winter, the wind swung suddenly to the south. The mercury shot up. Heavy rain fell, adding its water to the massive melting of the snow. By the next morning, Hampton Brook was rampaging over our lane, at the highest level we had ever seen it. We thought that perhaps the beaver dam had suddenly let go. But that afternoon, when we visited the spot, we found that, while a score of white cascades were tumbling over the barrier, only a small section, about five feet long and hardly a foot above the top of the stone wall, had been carried away. The reputation of the beavers as flood-control engineers was intact. And, as soon as the level of the pond dropped, this relatively small breach in the 125-foot dam was repaired.

Years ago Ernest Thompson Seton observed that no beaver dam is ever finished and no beaver dam is ever beyond the need of repair. In making repairs, the animals often use whatever comes to hand. They have employed stones, deer horns, even discarded beer cans. Among the Badlands of South Dakota, two dams were composed mostly of chunks of coal transported from a nearby river bluff. Some of the largest dams in the West are known to have been in use for a century and generations of beavers have contributed to their upkeep. At Trail Wood our dam is of the classic construction, its downstream side composed of short sticks anchored in the mud and tilted steeply against the pressure of the current.

Always the life of the colony depends upon the dam. It provides the necessary depth of water for the animals to find refuge from enemies, move freely under the ice in winter, store their submerged waterlogged food supply, and keep the entrances to their lodge well beneath the surface of the pond. Among their side effects, such barriers produce pools for fish, check erosion, reduce sediment in streams, provide flood control and, where ponds in time become filled with silt, create tracts of rich new soil. Throughout New England, some of the most fertile tillable fields mark the location of ancient silted-up and abandoned ponds made by beavers. Justly deserved is the title often bestowed on this builder of dams: the First Conservationist.

With the coming of spring the pond provided a whole new environment to the liking of many a wild creature—frog and turtle, dragonfly and wood duck. Hardly had the ice disappeared when we heard the trees resounding with the strange clacking chorus of wood frogs mating in the frigid water. When the sun climbed higher and the thermometer rose, green frogs multiplied in the shallows. Returning birds brought yellow warblers hopping among the sticks along the lower side of the dam and Louisiana water thrushes searching for food over the mud at the pond edge. Then came the whirligig beetles spinning across the dark surface of the water. And all around the pond, stranded sticks, gnawed clean of bark, grew more numerous. Our beaver-pond watching picked up as the days grew warmer.

About this time we discovered that, under the cover of darkness, the animals were building a series of baby dams, five or six feet long, across the swampy stretch below the main barrier. One had as its backbone a strong cable of wild grapevine that had looped down into the water between two trees. Each of these miniature dams held a pondlet impounded behind it. Day by day—or night by night—the number of these small barriers increased. When they extended in a zigzag chain across the lowland, I deposited flat rocks along their tops. These stepping stones provided a makeshift bridge to the other side.

In the deepening dusk, one evening in mid-June, I started along the winding course of this bridge the beavers built. Halfway across I stopped and looked around me. Light was draining from among the trees. Veeries and wood thrushes sang in the warm twilight. Fireflies winked among the ferns. And all around me, as I continued on, stepping carefully from stone to stone, the maze of little ponds glowed as they caught and held the last light from the unclouded sky. Night was coming to the woods—a clear night, a special night, the night of the full June moon.

On a hillside above the eastern edge of the pond, I groped my way to a small open space among thin bushes. There I settled down to wait. Below me the dark mirror of the outspread pond shone—like the pondlets of the lowland—with a luminous sheen reflected from the sky. While I waited for the rising of the moon, I sat listening to all the variety of the wild sounds of the far-from-silent night.

Clinging to bark on limbs and trunks around the wooded edge of the pond, gray tree frogs repeated the soft, musical fluttering of their calls. Their voices mingled together, rose and fell, went on and on. Once an airplane plodded across the sky high overhead. The faint faraway pulsations of sound coming down from its passage appeared to stimulate the tree frogs. Their wild chorus rose louder, filled the night, then died away to separate flutters of sound again. Green frogs along the pond edge, from time to time, broke out in sudden strumming crescendos. The abrupt sounds ran in waves, one increase in calling setting off another down the shore. Then all would fall silent as abruptly as they had become vocal. In the distance toward the north, a number of times, I caught a great squealing and flapping where some excitement gripped the wood ducks. This, like the strumming of the green frogs, would end as suddenly as it began. Then, a little later, a fresh outcry would commence. No breeze stirred the leaves around me and sounds carried far through the tranquil woods.

To the west, across the water, above the trees of the ridge beyond, the sky glowed with a pale greenish tinge. Looking behind me, I saw the rising of the full moon, its great disk of shining yellow lifting among the black silhouettes of the treetops. Minute by minute it continued its slow ascent, its misty light spreading more and more widely over the pond. In time, the lodge, the dam, the outspread water, the surrounding trees all lay bathed in moonlight. Those living sparks, the fireflies, went weaving in and out among the shadows, their little lights accentuated by the darkness, dimmed where the moonbeams were brightest.

All this—moonlight, frog calls, fireflies—has been, for ages, part of the nocturnal life of the beaver. Through untold generations these descendants of the extinct giant beaver of prehistoric times, an animal larger than a black bear, have been surrounded by such things. They were known to a million ponds when the first settlers arrived in the New World and they were part of the life of the animal colonies during those terrible decades of slaughter in the nineteenth century when, in the course of only twenty-four years, the Hudson's Bay Company sold on the London market nearly 3 million beaver pelts. Throughout Connecticut, a century ago, this largest of our rodents was reduced to a few animals making a last stand in the wilder parts of Windham Country, this far northeastern corner where we live.

Toward the direction of the lodge I heard a loud splash. The sound reverberated through the night. It jerked me to attention. But my sharpened gaze saw nothing except silver ripples running across the water in the moonlight. A little later, another splash came to my ears from somewhere at the northern end of the pond. Beavers were abroad. But they were extremely wary. I remained unmoving, almost without breathing, intent on the pond spread out in the mellow light below me.

Sitting there tense, expectant, waiting, I was overpowered by the feeling that I had experienced all this before. At first I thought I was recalling a time when Nellie and I lived in a trapper's cabin during a vacation in the Maine forest and spent hours observing a beaver pond from the slope of a wooded hillside. But my feeling went deeper. It seemed to return from farther back in time. Then I had it! Many years before, when I was small, I had read Charles G. D. Roberts' *The House in the Water*. And in reading it, I had lived with the boy in the story just such a time of waiting and suspense beside another, imaginary pond, remote in the wilderness.

Then I saw the beaver. Almost directly below me, its dripping head emerged into the full moonlight. Like the dark hub of an expanding wheel, it formed the center of an enlarging circle of shining ripple rings. Deliberately the animal examined the low mud barrier that extended like a wing dam from the end of the main structure. Water was streaming through a small gap. It was not a serious leak but it represented one of the many minor repairs a diligent beaver makes. While I watched its every movement, it scooped mud from the pond bottom, using both of its strong forepaws and holding the mass against its breast. Then this first of several loads was deposited in the gap. Included in the mud, I found on later examination, were binding materials, soggy masses of groundpine and other vegetation. As the beaver worked, fireflies drifted about it, their moving lights mirrored on the water.

A second beaver, announced by approaching ripples, came swimming to join the first. Perhaps I moved slightly to get a better view. Perhaps the moon, now much higher in the sky, shone more directly upon me. Whatever the explanation, the keen night vision of this animal detected my presence. A momentary stillness was shattered by the crack of its flat slapping tail and the splash of its dive. This was followed almost instantaneously by a second crack and splash as its companion disappeared. Then from farther away there came a third, and from different parts of the pond, in a chain reaction, a fourth, a fifth, a sixth. The surface of the water shivered. Ripples raced from all directions, met and overlapped. The sound of the flat tails meeting the water echoed from the side of the ridge. The beavers were gone. But the manner of their going had answered a question: How many beavers did we have? The number in our colony, according to the number of slaps coming from different parts of the pond, was six.

From then on the animals were active only out of my sight. The sounds they made came from the more distant portions of their pond. It was close to eleven o'clock. The chorus of the gray tree frogs that for so long had increased and decreased in volume like the rising and falling of a swing, began to fade away. Occasionally the deep trump of a bullfrog or the high barking of some distant farm dog broke the increasing stillness. Once during moments of unusual quiet, I heard what suggested chips being yanked from wood followed by a light rushing windy sound I could not identify. Had I heard a small aspen being felled somewhere in the night beside the pond?

Although, as long as I continued sitting on the hillside, I saw no more beavers, I came to realize that I was under constant surveillance. Once I flicked on and off my flashlight to check my wristwatch. On the instant the slap of a beaver's tail from a deeply shaded area of the pond set the colony on guard. And when I finally stood up, stiff from my hours of sitting, my movement once more triggered a chain reaction extending to all parts of the water. The first time, at the quick on-and-off flash of my light, only the watching beaver slapped its tail. The second time, when I made a movement and so apparently represented a greater source of danger, all the other animals had followed suit in rapid succession. Is there a difference in the initial warning? Does one kind of slap carry a greater sense of urgency and alarm than another?

I was pondering these questions when I started back across the stepping stones to find my way through the nighttime woods to the silent meadow, alight with fireflies, and the Trail Wood house rising white in the moonlight. The next morning when I opened my newspaper I learned that at almost precisely the time when I was watching the beavers repairing the dam, New England had been shaken by an earthquake. Although I was unaware of it, it had been powerful enough to register between 4.5 and 5.5 on the Richter scale. So during that memorable night beside the pond, I not only had seen beavers in the moonlight, I had seen them going about their ordinary ways in the midst of an earthquake.

CHAPTER SIXTEEN

A Naturalist in a Balloon

The last thing I expected to include in this book was a chapter on a balloon flight over Trail Wood. Yet here I am, beginning just such a chapter.

Many years ago I read a book by Florence Merriam Bailey about birdwatching in the Southwest. It bore the title *A-Birding on a Broncho*. In England W. H. Hudson wrote about his adventures among birds on a bicycle. But I recall no one who has written about birdwatching from a balloon, about nature as seen from the viewpoint of the wicker basket of a lighter-than-air craft drifting silently over fields and woods.

It was this adventure in seeing things from a new angle that tempted me when Charles MacArthur, of Tolland, Connecticut, wrote me about the observations he had made from his hot-air balloon and invited me to accompany him on a flight over Trail Wood at the time of the dawn chorus of the birds on a morning in May. Looking directly down on ponds, he wrote, he sometimes sees wild ducks taking off, a double row of dots left behind them by water dripping from their still-lowered legs. Once as he drifted slowly over the Tolland marshes, a gray male marsh hawk sailed back and forth, keeping just ahead of him. Another time, on a misty morning, a Canada goose, coming down for a landing, curved around the inflated balloon, craning its neck in curiosity as it circled by. During cross-country flights on winter days, tracks of wildlife on the snow-covered fields add to the interest. Several times, in this season of the year, MacArthur has floated above openings in the woods and looked down to see deer below him. In the stillness of the winter day he could hear their hoofs breaking through the crust of the snow as they walked.

Only in a balloon can such things be experienced. An airplane is too noisy, too fast to hear or see things close up. A helicopter is slower but its racket is deafening. A soaring plane is silent but it must remain well up in the sky to ride the rising air currents. Only the balloon can drift in still leisurely low-level flight just above the treetops. Only a balloon can provide the naturalist with an aerial grandstand seat for observing the natural world from this new and unfamiliar angle.

This viewpoint MacArthur has experienced on more than a thousand flights. He has ascended from Boston Common, drifted over the edge of the Everglades in Florida, floated above ranchland in Texas and looked down on Alaskan tundra in the first flight ever made in a hot-air balloon north of the Arctic Circle. He has given courses in the theory and practice of ballooning at universities as widely separated as in Wisconsin, Texas, and the state of Washington. Once, in South Dakota, he traveled forty-six miles before he landed. All these flights in varied places have been made without an accident. Only when the air is relatively calm, usually when the breeze is blowing no more than four miles an hour, does he go aloft. Such conditions prevail most often at dawn and sunset.

It was at dawn, on the last weekend of May, that we planned to take off from Trail Wood. The night before MacArthur, his wife, Anne, and two balloon enthusiasts, Richard E. Smith, of Hartford, and Dolly Hasinbiller, of Staten Island, slept in a huge camper parked beside our hemlock tree ready for daybreak preparations the next morning. This camper—with the balloon equipment stowed away at the rear—is, as MacArthur puts it, "the only aircraft carrier on the highways."

Before five A.M., the great envelope of nylon, copper-red on the outside and silver on the inside, was spread out on the grass. Two and a tenth miles of stitching, MacArthur explained, hold its 352 panels together. Inflated, the balloon towers seventy-four and a half feet high. It has a diameter of forty-six feet. And its 56,000 cubic feet of heated air can lift more than two tons. MacArthur had named his $4,000 lighter-than-air craft the *Henry David Thoreau*. Around the skirt at the bottom of the balloon ran black silhouettes of such endangered species of life on earth as the bald eagle, the kit fox, the blue whale—and man.

I watched with interest all the preparations for our voyage. The wicker basket which would carry us aloft sat to one side. In all the years of ballooning history nothing has been found superior to wicker construction. Wooden and metal gondolas have proved unsatisfactory. In a rough landing, they take the jolt all at once whereas the wicker construction absorbs the force of the impact in a rapid series of smaller jolts that reduce its effect. Two thousand feet of heavy rattan imported from Singapore had gone into the construction of the basket in which we would ride. In one corner three straps of heavy webbing anchored in place a metal tank of propane gas. The complete weight of the balloon, basket and tank, is 263 pounds. Trailing from the basket to a large metal ring lying on the grass—a ring that held the burner whose gas-fed flame would heat the balloon—I saw sixteen cables woven from strands of stainless steel. They support the basket, which dangles some ten feet below the envelope. The combined strength of these cables is sufficient to bear a load of thirty-two tons, or 64,000 pounds.

The dawn was bright, the sky clear. The air, I remember, was sweet with the perfume of the last of the apple blossoms. White petals from time to time drifted down, some to land on the nylon of the outspread envelope. The moment had come to inflate the balloon. Dolly volunteered to crawl inside the envelope and hold the fabric away from the flame of the burner to prevent damage during the initial stages. MacArthur ignited the pilot light, pointed the burner into the open mouth of the balloon, and increased the flow of gas. With a windy roar the heater, like a flamethrower, shot out a long tongue of yellow fire. In two minutes, he commented, the burner can supply enough heat to raise the temperature in an eight-room house to 150 degrees F. The balloon billowed up. MacArthur cut off the gas. Dolly, who had dropped to the bottom of the envelope where cool air was rushing out, came crawling through the opening. Among hot-air balloonists, the one who goes inside the envelope to hold up the fabric—no doubt because of what appears to be imminent danger of being roasted—is known as "the turkey."

The burner took up its steady roaring. This low-pitched sound of the rushing flame could be heard a quarter of a mile away. The immense bag of nylon lifted higher and higher, stood upright, swelled to the shape of a gigantic pear. It towered higher than the tops of the tallest trees. The upper part of its eastern side caught the rays of the rising sun while we stood in shadow. Looking up into the fully expanded balloon, the envelope appeared to us so thin and insubstantial we seemed about to ride a frail bubble in the air. Yet the nylon is strong, far beyond its needs, and the pressure on the envelope from within, even at the top, is only about four ounces to the square foot. Once, looking up as I was doing, MacArthur had seen against the light the silhouette of a pigeon that landed for a moment on the very warm perch at the top of the expanded balloon.

Each time the burner was turned off the nylon made a faint rustling sound, a kind of sighing like breeze among the leaves. With Smith gradually paying out a tether rope attached to the basket, MacArthur rose to treetop height for a short test flight. Then the *Henry David Thoreau* sank gently back to earth again. Nellie took her place in the basket and again the balloon ascended on a tethered flight. From a height of sixty feet or so she looked down on the treetops, on the white drift of the apple blossom petals, on the roof of the house, and, beyond the chimneys, toward the pond where slow-motion swirls of mist were rising above the water. Among the most dramatic scenes MacArthur recalls are those he has encountered on tethered flights when he has ascended at dawn through heavy ground fog and has emerged into sunshine with the white plain of the mist outspread, glowing and luminous, around him.

Again, in the almost windless air, Smith guided the balloon down to a gentle landing at the exact spot where it had taken off. I helped Nellie out and climbed into the basket. The wicker sides rose as high as my hips. While MacArthur tied the coiled-up tether rope to the side of the basket, I braced myself with one shoulder in the narrow inverted V formed by two of the supporting cables. It was a little after six A.M. The robin chorus was still at its height.

Then the voices of the birds were drowned out by the hissing roar of the long burst of flame that streamed upward into the gaping mouth of the balloon as MacArthur opened the gas valve. Slowly, gently our buoyancy increased. I hardly sensed the moment when we left the ground. There was no lurch, no sudden movement of starting, only a smooth transition from the immobile to the airborne. A hot-air balloon leaves the ground at minus four ounces. One hundred and eighty-four years after the Montgolfier brothers in France sent aloft the world's first hot-air balloon, we were lifting into free flight to drift for an hour wherever the movement of the air would take us.

Unhurriedly we mounted up and up. Yet all the time we seemed standing still while the apple trees, the lilacs in bloom beside the garage, the little knot of people looking up from our yard, the stone walls, the lane, the hickories, the paths our feet had made across the meadows drew back, the familiar scene opening out on all sides as it receded. I saw everything from a new perspective. I looked down on the open mouth of the fireplace chimney, seeing it as the swifts view it. I observed Firefly Meadow as the broad-winged hawks observe it as they soar on the summer thermals.

Rising above the trees to the east, the lifting sun cast the shadow of our balloon on the woods beyond the pond. With my eyes I followed its slow, stately sweep, saw it carried toward and then directly over my log writing cabin nestled among the aspens. Slowly we were gaining altitude. Slowly we were moving away toward the south, toward the valley of the Little River, toward the white spire of the village church. All across the valley the level rays of the early sun were probing among the mist of the treetops. When MacArthur cut off the heater, in the stillness—a stillness broken only by the low fluttering murmur of the pilot light and the occasional light metallic click of my camera shutter as I photographed the shimmering dawn scene below—we could hear crows cawing along the farther slope.

Then the dogs began to bark. All our neighbors seemed asleep but their dogs were wide awake. A chain reaction of excitement ran from farm to farm. Sometimes a balloonist can hear the barking of a dog as high as a mile above the earth. It is one of the sounds—along with the ringing of church bells and the whistling of locomotives—that carries highest into the sky. City dogs, MacArthur has noticed, are more excitable than country dogs. They keep on barking longer when a balloon goes over. The calmer country dogs bark for a while, then quiet down.

On one flight, he recalled, he reversed the sound of barking. It came from the balloon instead of from the ground. Early one morning as he drifted silently over an open field he spied a red fox hunting for mice in the grass below. He leaned over the side of the basket and bayed like a hound. The fox lifted its head. He bayed again. The animal looked around in all directions except up. When MacArthur bayed a third time, the fox trotted uneasily off toward the protection of a nearby wood.

We were now floating at an altitude of a little more than 200 feet. Ahead of us we heard the clatter of a small motor. One neighbor was up, running a cultivator along the rows of his kitchen garden. Until the shadow of the silent balloon passed over him, he was unaware of our presence. Then he looked up. The motor stopped abruptly. He went tearing into the house. Out poured his family, pulling on bathrobes. We saw them all shading their eyes, staring up at this apparition in the sky as we slowly drifted away.

MacArthur recalled a balloonist in Minnesota who made a flight during the hunting season. When he passed over one party of sportsmen, he saw them all aim their high-powered rifles at him. An uneasy moment followed until he realized they were simply looking through their telescopic sights to see him better.

Below us bushes and trees and fields and fences continued to move to the rear in a leisurely parade. Looking down, I noticed they were gradually rising toward us. MacArthur gave the balloon a short blast of heat to restore our buoyancy. There is always a lag of about thirty seconds before the effect of added heat becomes apparent. The skill of a balloonist is shown in his ability to maintain smooth flight, to avoid continually rising and descending while in the air. With experience he develops a sense of when to turn the heater on and off. He learns to anticipate the needs of his craft. On an average flight, the heater is on about one minute in twelve.

Between these bursts of flame, we drifted in such stillness that whenever we shifted position on the wooden floor the creaking of the wicker basket seemed loud in our ears. We could hear distinctly the faraway crowing of a rooster. Once when we passed over a small farm pond we looked down and caught sight of the balloon mirrored upside down on its dark surface. We saw the basket, the coil of rope, our peering faces, with the distended envelope of the red balloon beyond them. We seemed looking up at the craft drifting in the sky with a thin veil of cirrus clouds floating above it. As I was leaning over the basket's edge, looking down, making a note of what I saw, the stub of a black pencil with which I was writing slipped from my hand. I watched it as it fell down and down and down. It seemed to drift for a long time on its descent to the bushes and ferns beside the pond.

Up to now we had been moving down the western slope of the valley. Then the breeze left us. The parade of the landscape below came to a halt. We hung motionless in the air. A fence post directly below remained unmoving. We seemed anchored to it. Thus minute after minute we hung becalmed. Then the post began slowly to drift away toward the east. In the evening air flows down the slopes into a valley; in the morning it moves up the slopes again. Caught in this latter movement, we ascended the wooded side of the ridge toward the west. We were now traveling at right angles to our former course.

On one occasion, MacArthur found himself becalmed directly above a six-lane superhighway. Cars slowed and a major traffic tieup threatened before, trying different elevations, he found a breeze that carried him away. Ballooning is a sort of meteorological chess game. By constantly observing air conditions and studying the topography of the land, the balloonist decides on his moves. Sometimes the breeze just above the treetops is stronger than higher in the air. Sometimes adjoining layers of the atmosphere will be moving in opposite directions. In ascending from one such layer to another, MacArthur has seen the top of his balloon heel over as it entered air that was moving in another direction.

"One time," he recalled, "I traveled a mile and a half and then, at a different elevation, drifted back and landed at exactly the same spot where I took off."

On a balloon flight you do not start with a destination in mind. You are never going anywhere in particular. Your course is charted by the invisible flow of air currents. You become part of the breeze. All is leisurely; all is peaceful. There is no fighting the wind. A flight in a balloon, I discovered, has a drifting, dreamlike quality. You seem escaping from the nervous, violent, straining twentieth century. Throughout there is a restful sense of cooperating, instead of competing with nature.

We were about halfway to the top of the ridge when our progress halted once more. We hung for some time above the same wild cherry tree with a shining web of tent caterpillar silk among its branches. Then, in another right-angled turn, a new current of air bore us off toward the north. We had reversed completely our original direction. Thus the course traced by our flight suggested a giant horseshoe with one side longer than the other, a horseshoe with Trail Wood enclosed within it. Floating in a balloon, we seemed following the old New England custom of "walking the bounds" of the farm.

Our greatest height that morning was about 500 feet. Now, just above the treetops, we drifted across New Hill Road and the southern boundary of our land, over Seven Springs Swamp, Big Grapevine Trail, and the ancient white oak of the South Woods. Once we floated through a gap with the basket lower than the treetops on either side. Then I felt a momentary warmth on the back of my neck, flame shot up into the open mouth of the distended envelope and we rose to leapfrog in slow motion over higher trees ahead. The wicker basket of a balloon can drag without harm through the smaller branches and twigs of a treetop. In races in which the goal is to cover the least possible distance in a given time, balloonists sometimes drag their baskets through treetops intentionally to slow down their progress.

Looking into our South Woods as it passed below us, I caught glimpses of familiar trees—the lone white birch, the high, silvery swamp maple festooned with the coils of the ancient wild grapevine. Always before I had looked up at them, never down on them. Now for a first time I was seeing them all "from the other end."

Occasionally, from higher elevations, we had glimpsed birds darting from tree to tree or dashing across open spaces. We had caught snatches of their songs. For the dawn chorus of spring carries to a balloonist as much as half a mile above the earth. However it was here, drifting low above our own Trail Wood trees, that I was able to give myself up to the enjoyment of the springtime birds. Here they seemed most numerous of all. Here they were singing all around us —in the open spaces, on the woodland floor, amid the pale-green clouds of the fresh new foliage.

Ahead of us I could hear the watery song of the veeries of Veery Lane, the hoarse calling of the great crested fly-catcher that nested in the hollow of an old apple tree beside the pond, the bell notes of a wood thrush coming from the dusky ravine where the water of the perch spring pours from beneath the twisting roots of an ash tree. Again and again, the woods rang with the "Teacher! Teacher! Teacher!" of an ovenbird. For weeks I had been hearing many of these same singers. In some instances I could follow the voices down and picture their surroundings, even their perches and the very bushes and trees in which they nested.

Our course carried us directly above Juniper Hill. As we looked down on the rough green rings of the prostrate shrubs, a flicker, with its exultant call, rushed away down the slope in undulating flight, its white rump patch catching the sun, shining out against the dark background of the massed junipers. Rarely are birds alarmed by a balloon. It is too large, too slow moving, to be a bird of prey. It appears to pose no threat to them.

As we drifted north we skirted the western end of the pond. Going and coming, we reversed pond sides. Near Summerhouse Rock I saw a small flock of redwings and grackles scattered over the ground, gleaning cracked corn overlooked by the mallards. On and on a robin sang among the upper branches of an oak beside the little screened-in summerhouse, and from near its nest on the end of my log cabin among the aspens I could hear a phoebe untiringly call its name.

Tiny birds, brilliantly colored, dashed among the new foliage along Azalea Shore. May is warbler time. The voices of these lively little birds rang out below us—the loud "Wichity-Wichity" of the yellowthroat, the whiplash final emphasis of the chestnut-sided warbler, the small musical snore of the blue-winged warbler, the bright, ascending "Zee-Zee-Zee-Zee" of the prairie warbler. Their songs carried for a surprising distance. One of MacArthur's many plans is to float over the spring woods with a parabolic sound-collecting device pointed down beneath the basket of his balloon and thus record the dawn chorus of the birds as heard from the air.

Our slow drift carried us on, over the lowland path of Veery Lane, over the enchanter's nightshade growing under the leaning apple tree, out above the pasture where the woodcock ascends on its crepuscular song-flights in the early spring. Seen from the western side, the white Trail Wood house shone in the glowing misty back-lighting of the sunrise. I could hear the mewing of the catbird that nests in the lilac bush by the garage and the clear, musical "Che-wink!" of the towhee that feeds among the fallen leaves along the wall under the hickory trees.

Fluttering among the pasture flowers, as the Starfield passed beneath us, the small forms of white and yellow butterflies traced wandering paths. For nearly a mile on one flight, MacArthur recalled, he was once accompanied by a sulphur butterfly. It circled around and around his basket. Various other insects, including mosquitoes, have been encountered by balloonists as high as 200 feet above the ground.

Beyond the Starfield and Nighthawk Hill, riding over the woods to the north, we heard the voices of other towhees, of other wood thrushes and ovenbirds and veeries. Somewhere in the pasture we had left behind, a meadowlark sang of the "spring o' the year" and once, in a momentary lull in the bird sounds around us, I caught the faint jingling medley of bobolinks over a hayfield on the other side of Kenyon Road.

So far, for some reason, I had detected neither of those familiar woodland birds, the red-eyed vireo and the wood pewee. Now, just as we left the Old Colonial Road behind, I heard them both almost at the same time, the long unwinding monologue of a vireo ahead of us and the drawling notes of a pewee off to our right. Later, as we drifted on, my ears caught the voices of three vireos singing at once in different portions of the woods. Over one small stretch of swampy lowland, I listened to the fragment of an unfamiliar song, the voice of some bird I could not place. More than once dark robins dashed across little glades among the trees, and over and over again I heard the loud "slapping" call note of wood thrushes unseen among the undergrowth below.

We had begun rising slowly, extending our view ahead, looking for an opening in which to land. A balloonist always tries to descend before his tank of gas is empty, while he still has sufficient fuel to take care of emergencies. By now we had been aloft for nearly an hour and our supply of propane gas was running low. We leveled off at about 200 feet. Looking down from this elevation, I saw the spring woods below spreading away in all its delicate coloring, all its filmy clouds of tender leaves, all its special lighting at that hour of the dawn and that season of the year.

Only now, only in May, only in these days of spring and newly unfolded leaves, would we have encountered such a scene as lay outspread below us. To the balloonist sensitive to the beauty of the earth, each of the seasons brings its special memories—days of autumn foliage, dawns after nights of gentle rain when all the twigs of the treetops glitter with droplets of water, hours when winter snowfields are tinted by sunrise and sunset.

Less than an eighth of a mile ahead lay a long-abandoned field with woods closing in around it. Across its carpet of yellowed grass, scattered like low bushes, there were a dozen or so small wild cherry trees. Deer trails, one fresh, crisscrossed the opening.

"The rule in ballooning," MacArthur volunteered, "is to land in the first open space available when fuel is running low."

Our drifting balloon settled closer and closer to the treetops. We were advancing at a deliberate walking pace. As I looked ahead, I saw a dark-brown animal sitting up beside the entrance of a hole. I exclaimed:

"There's a woodchuck!"

It showed no signs of alarm as it sat watching our approach. So huge was the balloon and so leisurely was its progress that we were less than 100 yards away before the groundhog realized that we were drawing steadily closer. In a single movement it whirled and vanished in a headlong dive into the mouth of its burrow.

The reaction of animals to a landing balloon is a special field of study in itself. In farmyards pigs pile up at one side of a pen when a balloon drifts low above them. In a large pasture, Holstein cows move away like a school of fish when a balloon approaches in low-level flight. In one extensive meadow where MacArthur has landed on several occasions the horses galloped to the farthest corner the first time he descended but on later landings became so accustomed to the balloon they approached it in curiosity as soon as it came to rest. Deer in the woods are wary but not especially frightened by a silent low-drifting hot-air balloon. It is when the heater is turned on above them that its sudden roaring sound puts them to flight. Sometimes when a final short blast of heat is given a balloon to clear the last trees before descending into an open field, the tops of the long grass can be seen shaking in running lines where rabbits or woodchucks are racing away, frightened by the sound.

Losing buoyancy, the air slowly cooling in the envelope, we continued to descend. In landing, a balloonist controls only his vertical movement. His horizontal drift is determined wholly by the movement of the surrounding air. He may, on occasion, float helplessly along one side of a desired landing field and miss it entirely. On this morning we came in over the treetops headed squarely down the middle of the open space. In such a light breeze as ours, a balloon can land in an extremely small area. MacArthur once ended a flight in his own urban back yard. In descending, the force with which the basket meets the ground represents the sum of its forward and vertical motion. The strongest wind in which a landing can be made without dangerous dragging is one blowing about twenty-five miles an hour.

At the end of our flight, in the calm of the dawn, we touched the ground in the open glade almost as gently as a drifting soap bubble. Indolently the basket dragged through the top of the first wild cherry clump. I grabbed some of the small branches in an effort to hold us down but deliberately, irresistibly, they were pulled from my grasp. We touched again, rebound in slow motion six or eight feet into the air, settled once more, this time remaining stationary and upright. I looked at my wristwatch. We had been aloft an hour. The cost of the propane gas we had used in our flight was about two dollars.

For a time I remained in place, allowing the balloon to lose some of its buoyancy, to become "weighted." Then I climbed from the basket. Now, for the first time in an hour, I noticed that I felt the light breeze on my face. On all sides in the morning woods around us the chorus of the veeries, the wood thrushes, and the vireos rose and fell. Near and far, great crested flycatchers shouted hoarsely among the trees.

The opening in which we had landed, seemingly so wild and remote, was in reality only about 200 yards from Kenyon Road. To save labor in transporting the heavy basket and envelope to a waiting car, balloonists always try to land as near as possible to a highway. In a few minutes we heard the voices of those we had last seen staring up at us on our departure. They had followed our flight in MacArthur's big camper and now came trooping down the slope from the road through the scattered trees. With Dolly in the basket feeding the balloon just enough heat to keep it above the treetops and with Smith and MacArthur pulling on the tether rope, the *Henry David Thoreau* was "walked" to another open space close beside the road. There MacArthur pulled the ripcord. The great orange-red bag collapsed onto the grass. After it was rolled up and packed in the basket and the basket lashed to the rear of the camper, we all came back down Kenyon Road and up the lane to the house to eat a leisurely, if belated, breakfast.

As I ate, I thought back over the hour before. My dominant impression was one of unexpected tranquillity. Once the strangeness of the new conditions had worn off, this first flight in a balloon had been a time of tensions relaxed. There had been no sense of danger. We floated in calm and peaceful air. Our hour aloft appeared linked to no particular day or year; it seemed as timeless as the drift of spiderlings floating at the end of threads of gossamer through the sky.

And there had been an added fascination in that hour in the air. From a new adventurous viewpoint I had looked down as the crows look down on the hickory trees, as the great blue heron sees the pond, as the nighthawks in their whirling circus in the sky had beheld Nighthawk Hill. By the chance of the changing currents in the atmosphere, our journey had enclosed the woods and fields and brooks of Trail Wood. I had seen them spreading away below me from all points of the compass. There was where we had sat on the terrace listening to the whippoorwill in the first twilight. There was the site of my brushpile study. There the waterfall down which the wood ducks had tobogganed. There the Wild Apple Glade where I had found the family of white-footed mice in my hammock. There was the Old Woods Road along which we had paused to watch the little butterflies, the Dreamy Duskywings, spinning in the spring sunshine. Everywhere I had looked during that drifting hour, I had encountered strands of memory, threads of that web of recollections that bind together these Trail Wood years.

CHAPTER SEVENTEEN

Neighbors and Former Dwellers

"Nobody is ever going to build a monument to me," the hired man said. "So I will build a monument to myself."

And that is what he did. Set to picking up stones in a front pasture at Trail Wood, more than half a century ago, he piled them into a mound of flat gray rocks five feet tall and shaped like an old-fashioned skep, or straw beehive. The mound still stands on the highest point in the meadow beyond the brook to the east. You can catch sight of it between the trees as you drive up the lane. From the name of the hired man, it is known as Hughes' Monument. And from its presence the field derives its name: Monument Pasture.

Hughes was only one of a number of unusual people, doing unusual things, that have been associated with the history of this old farm. They have ranged from an owner who always drove high-stepping horses to another who brought his livestock into the living room and a vegetarian couple, inclined to nudism, "*very* peculiar but with hearts of gold." The earliest owner on record was the pioneer,

Thomas Grow, of Grow Hill. In 1801 he deeded the land on which our house stands to Andrew Durkee, grandson of John Durkee, who in 1715 brought his family from Gloucester, Massachusetts, and settled on "land on Little River." It was five years later that Andrew Durkee built the house in which we live.

Of all those sheltered beneath its roof and warmed by its fireplaces, during the more than a century and a half since, the one who is most often recalled is the eccentric poet, Andrew J. Rindge. His buckboard, drawn by an ancient sorrel horse, rolled along on wheels of different makes and sizes. It announced its coming by the squeal of ungreased axles. His buggy whip was a beanpole with a lash tied to one end. Heavily bearded, the poet rode with his head carried on one side. His poems, usually dealing with local happenings and the foibles and misdeeds of his neighbors, were tacked to the bulletin board in front of the village store.

In his youth Rindge, noticeably well dressed, was considered something of a local dandy. His first wife, Galista S. Fuller, was the beauty of Hampton. Only a few years after their marriage, at the age of twenty-seven, she died of tuberculosis. He married again, but his second wife died early of the same disease. From then on the poet's life went downhill. His heavy drinking, his unwashed dishes—surrounded by a cloud of flies—the sheep he brought indoors at lambing time, the quarrels with his neighbors over walls falling down and livestock running wild, all provided a kind of pre-soap-opera excitement for the village and a topic for endless conversations.

All during his latter life, Rindge lived alone. Without a family he treated his livestock as members of the household. Our long living room, with its great fireplace, is now floored with narrow boards instead of the wide boards originally there. The latter were so marked and scarred by the feet of his livestock they had to be replaced by a subsequent owner. In his bedroom he used to sleep with hens roosting around him. In what is now our bathroom he fed his pigs. It was his custom, in his old age, to boil a large kettleful of potatoes and then call in the animals. He ate some and they ate some.

Rindge's next-door neighbor on the north was Lyman Baker, a

thrifty Yankee bachelor. One of a family of fifteen children, he rode on a train and visited a large city only once in his life, when he went to Hartford to see the capitol. He was a saving and provident man who declared he would never eat another egg when the price rose above a penny an egg. As might be expected, he looked with disfavor on Rindge's shiftless ways, especially after the poet's walls fell into disrepair and his sheep wandered over into the lush grass of his best pasture. If they came over once more, he warned, he would keep the sheep. They did, he did, and a lawsuit followed. After lengthy wrangling and litigation, Rindge was awarded the animals. He celebrated the event by composing a long poem and posting it on the village bulletin board. It ended with the lines:

And now I thank the Lord, my maker,
I've got my sheep from Lyman Baker.

One summer day a horse trader of the old school drove up our lane. He said he had not seen our farm for nearly fifty years. Then, as a small boy, he had attended the auction following Andrew Rindge's death. He recalled that the man who bought his cows had spent days rounding them up. They were as wild and as wary of strangers as deer.

The end of Rindge's story came suddenly on a night of heavy rain. Returning from town, wearing a heavy overcoat that came almost to his ankles and with a bottle of whiskey in each pocket, he got off the train at the little station on the ridge and started for home. He had just turned into the lane from the road when he lost his footing and fell face downward in water collected in a low spot beside the way. There he was found in the morning. His hands were still in his pockets. He had drowned in only a few inches of water.

Contemporaries of Rindge included "Thunderstorm Bill," a man notorious for spitting when he talked, and the barefoot man who lived to be ninety-eight. On the first warm day of spring he always shed his shoes and not until frost had come in the fall did he put them on again. It was his belief, often repeated, that:

"There is strength in setting bare feet on bare earth."

Two others are remembered in the community for their words. One was the penny-conscious farmer who was fond of saying:

"I like to see it rain nights and Sundays so the hired man can get some rest."

The second was an aging countryman who passed a friend in a field as he was on his way to see a doctor because he thought he was losing his hearing. Coming home, he met his friend again and this exchange ensued:

"What did the doctor say?"

"I'll tell you what the doctor said. He said I had the dirtiest ears he'd ever seen in his life!"

Then there was the woman with whom everybody wanted to shake hands after church on Sunday. She always shook hands from side to side, as a dog wags its tail, instead of up and down.

Although the eccentrics may be fewer today, our village still seems to have more than its share of remarkable and unusual and interesting people. When we arrived, the community included the brother of a Hollywood motion-picture star and a woman who had had her cheek pinched by President Theodore Roosevelt when she was a baby in a baby carriage on a Washington street. A farm woman hired a babysitter so she could go out and shoot woodchucks. And one of our Hampton friends was told by a doctor that his sense of smell had been impaired by a virus disease. To him, skunks smelled like roses.

When I suggested to one woman in our village—who, in her eighties, was beginning to have trouble remembering—that it might be a good idea always to carry her name and address with her when she went to the city, her reply was:

"Oh, no. That will not be necessary. If I can't remember who I am or where I live, I will just go to a police station and tell them I am a lost Russian princess. Then my picture will be in the paper. Someone will recognize it and come and get me."

I remember one resident of the community who always spoke of "horrible katydids" and "terrible woodchucks" and "awful whip-poorwills" as though the adjectives were part of their names. Once, in pursuit of entomological knowledge, she asked me:

"What is the insect that is fat and brown and goes bang into the screens and that goes pop when you step on it?"

Another village friend of ours dislikes all butterflies.

"To me," she says, "they are just flying caterpillars, worms with wings."

Others come to mind: a kindly man who, when he was nearing ninety, had gravel spread at the edge of his pond when he found deer were slipping in the mud when they came to drink; another kindly man who took in a female dog that had been abandoned when neighbors moved away and was promptly presented with nine puppies—all females; a former resident who wanted his ashes scattered at a particularly beautiful bend of Little River; the woman in the hospital whose recovery was speeded by having her three cats smuggled in one at a time so she could see they were being well cared for; and the amateur auctioneer at a church sale in the village who held up a large, delicate cake and, as a special selling point, exclaimed:

"If you could only heft the weight of this cake!"

Originality and individuality are traits that are still apparent in Hampton. On one of my recent birthdays Helen Mathews sent me a greeting card she had made herself. It consisted of a fragment of wallpaper bearing this inscription:

A BIRTHDAY GREETING (SCOTCH)
This birthday card as you can see
Is nothing much at all.
'Tis just a piece of paper,
I pulled it off the wall.
But it carries bonnie wishes,
Tho' it didn't cost a cent,
For a birthday full of gladness
And a year of heart's content.

At the time when Evelyn Estabrooks was Postmaster at Hampton, and the post office was connected with the general store, she once appealed for new equipment by sending a poem to Washington. From the bureaucratic maze of the capital there came back a reply, also in poetry. Probably no one else in the world except another Hampton resident has the hobby of taking brides to their weddings in a vintage Rolls-Royce. So far, John Holt has transported twenty-five brides and their fathers to the church in this way.

Among all those we have heard about and known, among all the past and present dwellers in the village, I think the most remarkable was the blind naturalist, Annie Edmond. Even when she was past ninety, by feel and smell, she would identify wildflowers brought to her. By hearing she would identify birds. By feeling their bark and listening to the sound of the wind in their foliage, she would identify trees. Once when the Hampton Bird Club visited Brown Hill Pond, while other members scattered out hunting for birds, she remained in a car. There she not only piled up a list as long as that of all the others combined, but she reported hearing an additional bird, a woodcock, alighting in a neighboring low spot. A doubting Thomas walked through the area. A chunky bird flew up before him. It was a woodcock.

"When I was a girl on the farm," she once said to me, "what we now call wildlife we just called life."

The farm on which she was born in 1879 lies close to the source of Little River in the northern part of Hampton. She was nine years old when the blizzard of 1888 struck and she remembered vividly how great rollers, pulled by oxen, packed down the snow to open the roads for sleighs. Her memory for details of long-past events was particularly clear and accurate. This proved a great boon during the last twenty years of her life after unsuccessful eye operations had left her in darkness.

"In my life," she said, "I am so glad I stocked my mind with pleasant things to remember and think about."

Many of those things concerned nature. From time to time Nellie and I used to take her "olfactory bouquets" made up of sweetfern, spicebush, yarrow, catnip, sassafras, bayberry, and other leaves and plants whose fragrance would bring back recollections of her earlier days in the out-of-doors. Sometimes we would add a twig of black birch for her to chew for its wintergreen flavor. I remember one time when, after dark, we were lost on unfamiliar roads in the neighboring town of Chaplin. Fortunately Annie Edmond was along. From our serpentine progress, she recognized that we were following the top of a glacial esker and so was able to tell us where we were and how to get home.

When she first lost her sight, she told me, the thing that worried her was how she could tell when morning came. The robins beginning to sing at daybreak told her in the spring. At other times a clock with the face removed enabled her to determine the position of the hands by using her left thumb only. She cooked her own meals, using an electric stove. Her canned goods were marked in various ways. A rubber band around one can meant it contained peas, a paper stuck under the band on another meant it held peaches, and so forth. As her sister, Helen Mathews, who lived upstairs in the two-family house, put it:

"She was a spunky lady!"

During her latter years, before she died in the summer of 1971 at the age of nearly ninety-two, the multiplication of nature recordings, the sounds of bird song and frog chorus and insect music, helped her greatly, bringing nature indoors. As a result of her lifelong interest in flowers and ferns and birds and all the out-of-doors, she had stored up a vast body of observations and reminiscences. In those times when we sat talking beside our pond or riding through the country or listening to woodland sounds, we forgot that she was handicapped in any way. Always when we were in the company of this remarkable naturalist we felt we were receiving more than we gave.

Across Kenyon Road our nearest neighbor, Mabel Hetrick, was another strong-minded and unusual woman who reached her nineties. The widow of a physician, she stayed alone, took care of her house and herself and tended to her business affairs. Each Christmas she mailed out more than 300 cards, most of them accompanied by a personal note she often pounded out on the earliest model made of the Remington noiseless typewriter.

Her birthplace was Brooklyn, New York, where her grandfather had been a judge, her father an editor on *The Brooklyn Eagle*, her uncle the district attorney. For years she worked as a private secretary in an office on Wall Street where, on one occasion, she was put in charge of making all the arrangements for a tour of the United States by a former President of Mexico. It was almost below her office window that, on September 16, 1920, the anarchist's dynamite detonated in the Wall Street explosion that killed thirty persons, injured 100 more, and caused $2 million in damage. One of the other secretaries fell to the floor in a faint. Mabel poured a pitcher of water over her and brought her to.

"I am not the kind to faint!" was her succinct comment.

One Sunday after a sermon in which the minister spoke at length about how God had provided all the wonderful farming land for man's use, she asked him on the way out:

"If Adam had not sinned in the Garden of Eden, who would have helped him farm all that land?"

Not only did she keep her house in spotless order, but she was all prepared for atomic warfare. Stored in her basement were shelves of canned goods and rows of sealed bottles of pure water. In case of enemy attack the house across the way undoubtedly would have been the best prepared in Hampton. One evening, at the Little River Grange, she gave a talk on preparedness for atomic war, listing all the things that could be done. The only comment on her speech that she received came from a farmer who said as he was leaving:

"You know, Mrs. Hetrick, I can't do all those things. I've got my cows to milk."

Even to the end of her life her eyes remained particularly good. Across the fields in winter we could see her light burning each night as she read on and on, often until well after midnight. She never lost her interest in all that was going on around her. Only a few weeks before her death, when she was the oldest inhabitant of Hampton, almost ninety-three, she summed things up when she told me:

"I look to myself for companionship. I am never bored. I am never lonesome. I am interested in everything that happens. I am such *good* company for *me*!"

CHAPTER EIGHTEEN

Cabin in the Woods

In the chill of this early day in autumn, I light the first fire of the year in my Franklin stove. It fills the interior of my log cabin with the dancing light of flames, the snap and crackle of burning sticks, the faint fragrance of woodsmoke. Outside, blue-gray vapor trails from the chimney and drifts away up the slope of Lichen Ridge.

From the hammock the mice riddled, to the brushpile study the tree demolished, I came to this writing retreat among aspens and junipers above the far side of the pond. Here, within five minutes' walk of the house, I am surrounded by a scene that appears from my cabin windows as wild as though I were in northern Maine. Here I can read and write, uninterrupted by ringing telephones, more or less disconnected from the world. The chief sounds that I hear during the long summer days are the flutter of the aspen leaves and the calling of birds. Almost unconsciously, I take note of such small rustlings as the scurrying of a chipmunk, such snappings as that of a fly-catcher's bill as it snatches an insect from the air, such calling as that of goldfinches passing overhead. A brown thrasher smacks, a catbird mews.

This little house in the woods, about ten by fifteen feet, is constructed of pine logs and roofed over with shakes. From its eastern end extends a grape arbor formed of cedar poles and supporting the vines of red and white and purple grapes. To reach the cabin's rustic door, you follow the curve of an ascending path that leaves behind the weathered bridge over Stepping Stone Brook and the dark mass of a high red cedar tree. At different times in the spring, summer, and fall, as you ascend you find beside you pussywillows, bluets, dog violets, blue-eyed grass, and waves of goldenrod. If it is late April when you arrive at the end of your ascent, you discover shadbush in bloom and find the air fragrant with the perfume of its white flowers.

Even now, half a dozen years after the cabin was completed, when I unlock and push open the slab-covered door I am met by the clean wood smell of the interior. Some of my most pleasant hours have been spent here on days of gentle summer rain. Then to the smell of the logs is added the scent of the shower on the dry dust and the redolence of the wet aspen leaves coming through the open door. Soothed by the rhythm of rain on the roof, breathing moist air scented with the perfumes of the earth, I relax as the dusty leaf relaxes in the downpour. Once when I was looking out the doorway, I was surprised to see a black swallowtail butterfly winging its way through the falling rain, pursuing its wandering course across the dripping landscape.

It is among country folk that rainy days are really appreciated. In rural areas rain comes into its own. In the city, where most radio and TV weather reports originate, the weather is "good" when the sun shines, "bad" when it rains. What the city man invariably wants to hear is a report of sunny weather. A shower is an annoyance, an all-day rain a disaster. Stay away rain! Yet to say "May all your hours be sunny" is another way of saying "May you perish in the drought."

As my eyes wander about the interior of my writing cabin, they encounter the broom, dustpan, poker, and shovel behind the open-faced Franklin stove; the rustic rocker where I read; the straight-backed chair beside a folding table—a table inherited from a forest camp in

Maine—where I write; the lightweight typewriter that traveled with us through all the seasons in America and the pack basket made long ago by our son, David. It rests on the floor beneath the rear window of the cabin, a window that looks toward the south into the heart of the aspen grove.

After the pack basket had been there for several years, I shifted it one late-June day to a new position on the floor. In lifting it, I uncovered a mystery that, to this day, has never been completely solved. I noticed what appeared to be a mass of black shiny seeds. Bending down for a closer examination I saw that each seed was the severed head of a carpenter ant. There must have been at least a teacupful of these heads. How had they arrived in this hidden place? The first thought that flashed through my head was that I had stumbled upon a battlefield of the ants. Heads such as these are hard and virtually indigestible. Long after the rest of the insect has been consumed in carnivorous plants, the head of a carpenter ant will remain intact.

A second hypothesis occurred to me. Could some predator have caught the ants, fed upon them, and then discarded the inedible heads beneath the pack basket? A third explanation, although I have never been able to prove it completely, now seems most likely. This third suggestion is that I had uncovered the kitchen midden of a colony of carpenter ants; that after consuming the softer, edible parts of members of the colony that died, the survivors had deposited the indigestible heads in this secret hiding place beneath the pack basket.

The ways of the ants and the ways of insects that ants prey upon represents an especially fascinating area of natural-history investigation. Not long ago scientists studying a neotropical wasp, *Mischocyttorus drewseni*, observed that the greatest hazards faced by these insects were raids made upon their nests by ants. But the wasps—or rather the long processes of evolution—had provided an effective safeguard against such forays. Within the bodies of the females glands produce a viscous fluid. When this is applied in a heavy coating over the long stem of the nest, it turns the invaders away, for the glandular secretion contains an ant repellent.

How did it take place that this particular chemical should be produced within the bodies of these particular wasps? How did it happen that leeches possess a chemical in their sucking mouths that prevents coagulation and increases the flow of blood while deadly serpents inject with their venom a chemical that stimulates coagulation, thus locking the poison within the victim's body? How did the little water beetle, *Stenodus*, mentioned in the chapter on our pond, evolve the glandular secretion that destroys the water film behind it? Such instances of glands appearing just where their specialized chemicals are needed most, where the lives and welfare of their possessors depend upon their presence, represent for me the most baffling aspect of evolution.

But to return to my writing cabin. When I look up from my work, I see, ranged on the walls around me, mementos of earlier times, relics of past adventures. Hanging there are the rubber flippers that propelled me down among the eelgrass of Shinnecock Bay in the fall; snowshoes that carried me over the drifts of the deer yard in northern Maine in winter. On a nail there hangs a gray cap with "The Black Rail" printed in ink on the visor. This is the cap I wore when I saw that rare little marsh bird for a first time on the sea meadows of Cape May. Near the stove there is a rusted rapier dug from a swamp in northern Indiana and in one corner of the cabin a piece of picturesque driftwood picked up on the rocky shore of Crocker Lake, near the Canadian line in Maine. And there, beside the door, is the worn insect-collecting net that once belonged to William T. Davis, the gentle Staten Island naturalist, authority on American cicadas and author of that perennially engaging book of his outdoor reminiscences, his *Days Afield on Staten Island*.

During his later years the handle of the net—well polished by use—doubled as a cane on his field trips. If you examine the long ferrule of brass which attaches the handle to the net, you will see that it is heavily dented. When he was nearing seventy, this mild-mannered collector of insects was following a lonely road on Staten Island with his net in his hand when he encountered three young thugs who blocked his way. One pulled out a gun and demanded his money. This so infuriated the naturalist, who had spent a lifetime helping and being considerate to other people, that he swung his net, whacked the thug over the head with the brass ferrule, then pursued the three down the road flailing them with his butterfly net.

Later a friend asked him:

"But weren't you taking an awful chance?"

"I suppose I was," he replied. "I didn't think of it at the time. But I might have injured them severely."

Other mementos of various times and places cover the top of an unpainted bookcase and a long shelf that extends across one end of the cabin beneath three of Edward Shenton's original drawings for *Dune Boy*. They include colored rocks from Death Valley, a piece of petrified wood from the Badlands of South Dakota, the cone of a digger pine from Sutter's Mill in California, a gastrolith, or stomach stone, from a prehistoric dinosaur, a small bone from an extinct great auk, and a fragment of Indian pottery John Muir picked up while traveling with John Burroughs through the Painted Desert in Arizona. The skull of a kit fox, weathered to alabaster whiteness, came from the California Desert; the horn of a young bull elk was shed above timberline in a mountain meadow 11,000 feet up in the Colorado Rockies.

In a far corner of the long shelf there rests an old-fashioned kerosene lamp. It sheds its yellow light on my pages on those rare occasions when I work in the cabin after dark. Close by its illumination seems adequate enough. But how dim it really is I discovered one evening when I returned to the house for something I had forgotten. On the slope rising beyond the pond I paused and looked back expecting to see the yellow rectangles of lighted cabin windows standing out in the darkness. Instead, to my amazement, I saw nothing of the kind. With the lamp burning in one corner, the interior of the building was so poorly lighted it appeared to be in darkness.

There are four windows in my cabin. Each, viewed from the inside, frames a different wild scene. One takes in the path, the end of the pond, and part of Firefly Meadow. Another looks down on the tangle of Azalea Shore—that mingling of ferns, juniper, shadbush, wild cherry, swamp azalea, high-bush blueberry, cedar, and the old apple tree overrun by wild grapevines. A third faces toward the west. Through it your eyes sweep across an expanse of goldenrod, gray birch, and sumac to the rise of Juniper Hill and the oaks and maples that crown the ridge beyond. But dominating this scene is the most unusual feature of the cabin's surroundings. This is a clump of prostrate juniper that has spread out from the center in a living mat of green that has expanded year by year. It now forms a circular mass with a diameter of nearly forty feet.

The last of the four windows, cut in the back wall of the cabin, has already been mentioned. It looks to the south and into the interior of the aspen grove. Of all the trees in North America, the aspen has the widest range. It flourishes from northwestern Mexico as far north as Labrador. Among the aspens that grow behind my cabin I see, in winter, grouse getting buds; in spring catkins clustering on the twigs as on pussywillows; after late-summer rains mushrooms of varied colors springing up amid the smooth-barked trunks. Every autumn a feature of the season is the dry scratching of the falling leaves as they strike and slide down the six small panes of this southward-looking window.

Just outside this window, from among the branches of the aspens, a red squirrel sometimes catches sight of me through the glass. As long as I remain still, it watches me in silent concentration. But if I make the slightest movement, it goes into a sputtering, barking, coughing paroxysm of indignation. It and its kind, it seems to say, held squatter's right to this hillside long before I was born. It is *I* who am the interloper.

In contrast the woodchuck that has its burrow ten or fifteen feet back of the cabin among the trees appears more placid and philosophical. Stealing to the window, I occasionally watch it as it sits quietly beside the entrance of its hole. A second entrance, providing a way of escape in an emergency, is hidden amid plants farther back among the trees. I remember one late-September day when it lay flattened out on the mound of excavated earth. Its body was limp. Its eyes drooped. It nodded, half asleep. It was close to the threshold of hibernation. One thing I have discovered in watching this animal at times when it is alert and wide awake. When it looks about it intently, as when sensing danger, it can go far longer without blinking than I can. In one instance, I blinked four times to its once.

On the four shelves of the bookcase in my cabin there is ranged an odd assortment. In the main the volumes are old favorites that I dip into from time to time for pleasure—Amiel's *Journal*, Thoreau's *Walden*, *The Rubáiyát of Omar Khayyám*, *Alice in Wonderland*, Emerson's *Essays, Robinson Crusoe*, Major's *The Bears of Blue River*, Seton's *Two Little Savages*, and the *Collected Poems* of Thomas Hardy. Books have a way of drifting back and forth between the shelves of this little open bookcase and the large working library that extends around three walls of my study at home. One of the strangest volumes to make that journey was published in Wallingford, Connecticut, in 1875. It bears the title of *Footnotes, or Walking as a Fine Art*. Frankly it is a pretty dull book. Yet the author says in his foreword that all during the time he was writing it he was sure the spirit of Henry Thoreau was guiding his pen.

Anyone who sits silent and motionless—reading or thinking or writing—appears the least alarming of humans to wild creatures. So, in retrospect, memories of reading at my cabin and memories of wildlife merge. Once, on a summer afternoon, as I sat in my rocker before the open door, nearing the end of Richard Jefferies' *The Story of My Heart*, I looked up to see a cottontail rabbit in the doorway. It was hardly four feet away. It had its delicate little forepaws on the sill and was regarding me steadily with its large soft brown eyes. Even so slight a movement as lifting my head revived its natural timidity. In three bounds it disappeared among the clumps of juniper. On a later day, when I was sitting in the same chair in the same place, beginning *Goethe's Conversations with Eckermann*, I caught sight—out of the corner of my eye—of a red squirrel approaching in a series of little leaps. It ascended to the top step and for nearly a minute, while I remained frozen in position, looked curiously in at the open door. Familiar with its exterior, the wild creatures seemed inquisitive about the interior of the cabin.

Almost as soon as it was completed, this log structure was accepted as part of the landscape by the wildlife living around it. A chipmunk promptly dug its hole under the stone steps. A brown thrasher used the ridgepole as its singing perch at sunset. Flycatchers dropped beetle shards and butterfly wings on the steps. The woodchuck excavated a second burrow under the foundation, providing itself with an auxiliary home, a kind of basement in miniature beneath my cabin floor.

From time to time deer leave their heart-shaped tracks beside the building and along the path descending to the pond. One of the favorite dusting places of the grouse is beneath an overarching clump of juniper just outside the cabin door. Mason wasps, each summer, cement their nests of mud to the wood of the window frames. A pair of robins once nested above the door and, year after year, phoebes have attached the cup of their moss-covered nest to one of the upper logs where it is protected by the overhang of the roof.

When I sit in the doorway or on the steps just outside, I seem —like the cabin itself—to merge with my surroundings. Crows caw as they fly by but most of the other inhabitants of the hillside pay slight attention to me. Undoubtedly they know that I am there. But as long as I remain motionless, they accept me as an inoffensive feature of the landscape. I see metallic-green tiger beetles run on nimble legs in swift dashes over the steps below me. I hear bobwhite quail give their loud clear rallying call on the ridge behind me. I watch wood-boring carpenter bees drone past me to fill the cabin with their heavy humming as they wander about the interior.

At various times when I have finished reading and have been sitting on the cabin steps with a book on my knee, I have been diverted by particular occurrences around me. Once it was chimney swifts dashing against the dry twigs of a dead branch high in a wild cherry tree beside the pond. The birds were breaking off the outer tips for use in the construction of their nests. Another time, early on a morning in fall, my ears caught the ragged clamor of approaching geese. Heading south, a long V of flying birds passed directly over the cabin. They were spotlighted in the low rays of the early sun. Their contrasting plumages, snow-white bodies, inky-black wing tips, stood out brilliantly against the clear blue of the cloudless sky. These forty-five migrating birds added the hundred and thirty-third species to the Trail Wood list. Only on this one occasion have we seen snow geese crossing our skies.

Memories such as these weave together my cabin days during the spring and summer and early fall. As autumn advances, as the dawn comes later and the dusk earlier, as the chill of the season increases, I come less often to this little world of my hillside. But from time to time, well on into the shortening days, I continue to read beside the Franklin stove. Then the cold clamps down. Snow whitens the aspen grove. With the cabin door locked for a final time until spring I bring home the old rocker to use beside the fireplace, part of the way sliding it along over the snow like a sled on its runners.

CHAPTER NINETEEN

White Winter

With a pocketful of ripe fruit, round and red, crisp and filled with tart, refreshing juices, I turned away from the wild cranberry bog above our beaver pond and started home over the snow-clad hills. The sun shone from a cloudless sky and the weather was mild on that December morning.

Halfway up the first slope I halted. My footsteps, where I had come down the hill a quarter of an hour before, stood out dark gray, almost black, imprinted on the whiteness of the snow. I quit munching on the wild cranberries. Leaning close, I saw the bottom of each depression was covered with what appeared to be dark dust or a dense sprinkling of black pepper. As I watched, I perceived the multitudinous particles were all constantly in motion. Living dust had collected in my footsteps. Each of the specks was a minute, almost microscopic insect, a springtail or snow flea. On this thawing day in December, they had come in incalculable numbers to the surface of the snow.

Everywhere I looked around me on the slope I saw a thin drifting of this same animate dust. In Lilliputian leaps the dark particles were exploding into the air. Beneath the tails of these tiny creatures, the microscope reveals, there is a stiff hairlike projection that can be folded, under tension, and locked in place. When it is released suddenly, the snow flea is catapulted into the air. Thus it moves about as though on a pogo stick, arcing above the snow in a succession of explosive little hops. It rises into the air without wings. It jumps like a flea without possessing jumping legs. Its own six short threadlike legs are employed only in laborious crawling.

I paced off the area where the springtails were thickest. I found it measured twenty feet in width and forty-five feet in length. In those 900 square feet, perhaps hundreds of thousands of these minute insects were scattered over the snow. In many places at Trail Wood—on Nighthawk Hill and across the Starfield, around my log cabin among the aspens, beside Fern Brook Trail and amid the snow-covered clumps of Juniper Hill—we have encountered these curious creatures that have crept upward into the sunshine during days of winter thaw. But nowhere else have I seen them concentrated in such numbers over so wide an area as here on this northward-sloping hillside above our cranberry bog. For a time I toyed with the idea of tracing "TRAIL WOOD" in the soft material and watching the letters grow darker as the words, written in springtails, became steadily more distinct.

These primitive insects belong to the *Collembola*, a small order comprising about 1,200 species. They are found throughout the world. But they appear in greatest abundance in the temperate zones. Most forms depend on decaying vegetable matter for their food. But a few are predacious and carnivorous. Others emit a faint glowing light. Still others wander over the surface of standing water, feeding on diatoms and algae. Toward the end of one winter, when the ice began to retreat in Whippoorwill Cove, we discovered the surface of the cold water was swarming with an aquatic form of the springtails. Thousands, hopping about on the invisible surface film of the cove, had collected along the edge of the ice. Their bodies, congregated in a dark stripe, outlined each curve at the water's edge.

The species of springtails that appear on the snow of our fields spend most of their lives hidden under stones and litter, in moss, in the crevices of bark, amid the leaves of perennial plants, even in the nests of ants. During one winter, in January, the Wolf Month of the Anglo-Saxons, I dug down through the snow to the rosette of a mullein plant. I brought it home in a paper bag. Sitting that afternoon beside the fireplace, I made a census of the small creatures that were hibernating between the woolly leaves as though between thick soft blankets. My magnifying glass revealed a minute spider and a number of tiny brownish beetles. But the real population of the mullein plant consisted of springtails. I counted eighty-two snugly protected within the hibernaculum of this one rosette.

A letter reached me, some years ago, from western Massachusetts. As a small girl on a farm, the writer recalled, she used to see snow fleas every winter. But it had been a long time since she had noticed any. She had difficulty making people in the town where she lived believe there ever had been such things. Had springtails, she wanted to know, become extinct? Because they are so tiny, because they are so secretive, the average person never sees them unless he happens to be abroad in the country during winter thaws. Surprisingly little is known about their lives. I have never found even a definite explanation of why they congregate on the snow. Presumably it is a mass mating performance. E. O. Essig, in his widely used *College Entomology*, gives the simple explanation for our lack of knowledge of the life of the snow flea. "For the most part," he notes, "the biology of these insects has received little attention."

On the road to winter, each year, there are many milestones. There is the fall of the leaves, the arrival of the first junco, the first frost, the first ice on the pond. But overshadowing all the others is the coming of the snow. Normally our first snowfall arrives around Thanksgiving but some years we reach Christmas with the fields still open. Whenever they come, those first falling flakes stir something deep-seated and primitive in our natures. They bring a sense of mingled exhilaration and foreboding. We feel the urge to take stock of our resources, batten down our hatches, get ready for a testing time ahead.

We hurry to the woods—as though for a last time—and come home laden with dry sticks for fireplace kindling. I pile up logs on the hearth. I check the fastenings on the shed doors, look at the level of the fuel oil in the tanks in the basement, go over our stores of food in the pantry. Nellie lines up our winter coats and boots and caps. We prepare as for another blizzard of 1888. Then the flurry of activity subsides. We relax and settle down to our enjoyment of the white New England winter.

In succeeding days we watch the snow descending in varied ways. We see it drifting down from a windless sky; riding on gale winds out of the north; descending in a blinding curtain of white through which I hear the rumble—like the sound of a train passing over a trestle—where the highway plows are out in the storm. We observe large flakes sinking slowly, hanging in the air, buoyed up like thistledown; small hard flakes spattering down as though hurled by the handful; light and tiny flakes sifting in a fine shimmer from the sky. Often our winter snowfalls are preceded by a windless waiting time, a sense of listening, of expectation everywhere. At other times, the storm pounces, arrives with a rush, and the fall of snow is heavy from the beginning. Still other snowfalls alter as they continue. I remember one daylong changing storm that brought snowflakes in a succession of types: dry and granular, large and downy, finally tiny sticks or rods of ice—such needle snow as forms the building stuff of cirrus clouds.

Following one winter gale that ended in a fall of coarse granular snow, Nellie and I tramped through the North Woods. There we made an interesting discovery. Throughout our region birch trees are common. In the succession of forest trees they are one of the earliest to appear. On this morning, the surface of the snow mantling all the woodland hills was pebbled by the granular character of its topmost layer. Over acre after acre in each depression we found a snuff-brown particle, one of the fine windblown seeds of the birches. The gale that had pounded the woods during the night had scattered them across the slopes. The character of the snow had provided a natural mechanism for their even distribution. All the small depressions were almost equidistant from each other. Later the melting of the snow would lower the seeds to the ground, dispersing them uniformly over a wide area.

Snowfall of a different type resulted in another winter adventure of ours. It was on a day when the fallen flakes were so light, so powdery, that each time a bluejay took off two small white puffs shot up from the tips of its downbeating wings. The wind was unstable, the weather changeable. Just as I glanced toward Monument Pasture, my eye was caught by a shining funnel of white. High, whirling, wraithlike, it rose above the hillside. As we followed it with our eyes, it advanced up the slope and passed over the brow of the hill. In this white whirlwind we were seeing our first snow devil, that rare winter counterpart of the brown dust devils that go spinning across plowed or dusty open fields in summer.

In his winter journals Henry Thoreau often refers to "finger-cold days." On the other side of the Atlantic, Gilbert White speaks in his Selborne journal of snow "half-shoe deep." Usually our snows are deeper. But the only kind that keeps us housebound is the sort that extends across our fields after the passing of an ice storm. Then the snow layer is topped with a glittering crust that supports us part of the time and lets us break through suddenly the rest of the time. This lurching progress is the most exhausting knee-wrenching travel I know.

When we first came to Hampton, we looked at our long lane and wondered how often we would be snowed in in winter. As it turned out, some years we have been plowed out only three or four times, once only twice, from fall till spring. During other winters of greater snowfall we have had to be plowed out eight or ten times. Once we were marooned from Sunday to Friday. To get the mail I went down the lane over the great drifts on snowshoes. Both of us were well and those days of simple routine, without distractions, stand out in memory as particularly happy ones.

During the abnormal month of February 1969, when storm followed storm and a greater amount of snow fell on Connecticut than in any other February in forty-six years, we were snowbound for a total of nine days, close to one-third of the whole month. One of those

February storms, according to the weather bureau in Hartford, dumped 600 million tons of snow on the state. While we were snug in our white house in the white storm, 23,400 tons of snowflakes descended on Trail Wood.

The marks of my snowshoes, left on the surface of the drifts, are the largest tracks we see in winter. But many others, tracks of the winged and tracks of the four-footed, are being imprinted night and day on the white expanse of the snowfields. Reading the stories they record adds to the interest of every walk in winter. Lacelike trails of white-footed mice join the juniper clumps together. The delicate imprint of the small wing feathers of a tree sparrow marks the place where it darted into the air from beside a dry goldenrod. Along the edge of the pond the dragging tracks of a crow record its search for food among the grass clumps. And down the length of Azalea Shore we see where a mink has bounded along, abruptly changing its direction, suddenly reversing its course. This stilled record of former activity remains remarkably alive; the very footprints seem charged with some of the animal's intense nervous energy.

Two other tracks have made lasting impressions on my mind. One was a charming sequence of prints running along Fern Brook Trail. After a light snow, two or three inches deep, Nellie followed this path. The next day we followed it again. During the night a young deer had come this way and in each of Nellie's footprints there was the heart-shaped mark of a small, delicate hoof. The animal had stepped exactly where she had walked. She had broken trail for a fawn.

The other recollection concerns the track of a fox. We noticed it on a morning when the thermometer stood at five degrees below zero. Running in a straight line up the lane, the pawprints stood out strangely in the morning light. Instead of being imprinted in the snow, they were embossed on its surface. At each step the weight of the animal had pressed down the soft snow of the drifts, compacting and solidifying it. Then the wind had risen in the night. By dawn it had scoured away the loose snow, leaving the more solid areas of the pawprints slightly raised above the surface.

Add to these tracks the tracks of the wind itself. At no other time of year is the movement of the invisible air so apparent as in the white months of winter. We see it recorded in blowing clouds of surface flakes, in solidified ripples and waves, in sweeping lines, and in the streamlining of the drifted snow. It adds, in its immaculate sculpturing, its own special contribution to the beauty of the winter landscape. Blizzards and gales driving out of the north—the kind of weather that makes headlines in the Florida papers—leave our scene transformed. Downwind from every bush and tree and boulder and old stone wall the snow is scoured into smooth and flowing lines. Automotive designers in Detroit could learn much from the molded sweep of its streamlining. In the dawn after such a storm everywhere we look there are new forms of plastic beauty glowing and glistening in the early morning light.

Primarily, I suppose, the snow months of winter, the so-called dead season of the year, is the time of the young and strong. As people grow older the snow—the joyous snow of our youth—seems more and more the enemy. A barometer of our health and spirits and strength is our enjoyment of the white winter. But even when the exhilaration of plowing through the snow in high-topped boots, the exultation of striding over drifts on snowshoes, is something of the past, the beauty of the snow remains to be enjoyed.

For in it we find a double beauty—the beauty of form and action and the beauty of light and color. Almost as ethereal and evanescent as the changing illumination of an aurora are the subtle tints that play across the surface of the snow. We see it tinged with delicate hues and pale shadings, with faint greens and reds and lavenders, with tintings of salmon and lilac. At sunset and in the blue ebbing light that follows, we often end our days in winter by watching minute after minute the sequence of hues that glow and fade on the snow-clad fields toward the west. On nights of the full moon, the drifts shine with a luminous silver light while shadows under the bare trees sprawl in somber tarnished gray. Other shadows, cast by the long rays of the sinking sun toward the end of short winter afternoons, extend in violet-blue over the whiteness of the snow. E. B. White once wrote: "I am always humbled by the infinite ingenuity of the Lord, who can make a red barn cast a blue shadow." How varied are the colors of "snow-white" snow and how much they add to our enjoyment of winter!

For several memorable minutes, one January day in 1966, colors of a different kind glowed over the snowfields beyond Hampton Brook. The weather that day was changeable. Snow, sunshine, freezing rain alternated through the afternoon. For a short time, a little after four o'clock, the sun broke free from the clouds in the west and threw the brilliance of its rays on a shower of cold rain falling in the east. Looking that way we saw, high above the white slope of Monument Pasture, one of the rarest forms of tinted winter beauty. The colors of a rainbow's arch glowed over the snowfields, strengthened, then gradually faded away.

What is a snowstorm? Webster's simple definition is: "A storm of falling snow." But a country definition here goes farther. It isn't really a snowstorm unless the deposit of snow is deep enough "to track a cat in." Why, generations ago, were winters more severe than now? The explanation offered hereabouts—incidentally a belief held more than 2,000 years ago and one I have found recorded in Edward Gibbon's *The Decline and Fall of the Roman Empire*—is that there were greater areas of forests then where snow remained unmelted later in the spring, thus lowering the temperature of the climate. Contradicting this idea is the fact that the area of woodland in our state is greater today than it was a century and a half ago. Abandoned farms have been overrun by trees.

During the fifteen years since we came to Trail Wood we have experienced all the winters here with a single exception, the one we missed while traveling through the winter from coast to coast. These rugged seasons have rounded out the adventure of the year. During them we have heard all the voices of the cold: the icy whisper of sleet striking on dry leaves, the creaking of frosty tree limbs, the squeak and squeal of snow under our feet on sub-zero days. We have seen the December sunshine and the subdued grays of the leafless trees. Where else do grays reveal the richness of variety found in a winter woods? Such things as these weave together the fabric of our memories of these months of colder days.

All the while we were aware we were seeing beauty, hearing sounds, experiencing things that the hibernator in its long sleep, the migrant far to the south, would never know. How much we would regret losing all the things that they miss! Everyone looks forward to the spring. And so do we. But we want each season in its turn. We want all the changes of the weather. We want the year complete. So, at Trail Wood, for a decade and a half, we have found in this rugged time of cold, this white winter, this period of tinted snow and fireplace fires, the winter of our content.

CHAPTER TWENTY

Visitors from the Woods

Almost every evening when the sun goes down and the long winter night begins, the gray fox comes out of the woods, over the snowy fields, to feed on scraps outside our kitchen door.

Sometimes it arrives in the dark of the moon, materializing suddenly in the light falling from the window. Sometimes it comes when the moon is full and shining from a clear cold sky. We watch this animal, so ancient in folklore and legend, running about effortlessly, tirelessly, over the drifts. Its every movement has an airy lightness. It appears to float over the snow. All its motions are graceful. On such occasions a wild fox coursing among the drifts in the moonlight appears a vision of life, of warmth, of movement amid the still, crystalline world around it. It is something we look forward to when the white months of winter arrive.

On Christmas Eve, one year, when the moon was full, we turned out all the lights in the house and for a magic quarter of an hour watched a fox wandering over the glistening snow in the silver mellow light. It trotted this way and that, its moon shadow slipping over the drifts beside it. Once it stopped and stretched itself like a dog. A little later it halted suddenly and vigorously scratched its left shoulder with its left hind leg. A flea was also alive and well in the winter night. We saw the fox leap to the top of a wall and leap down again as lightly as a ballet dancer. Its flowing bushy tail added to its fluid grace in motion. A little later it took a running start and scrambled part way up the slanting trunk of an apple tree to reach a piece of suet wedged in a lower branch. The gray is sometimes called the tree-climbing fox. On occasion it ascends upright trunks like a boy shinnying up a tree, hugging the trunk with its forelegs and hitching itself upward by pushing with its hind legs. Later it backs down again. In the tree it may leap like a cat from branch to branch. Once a gray fox was observed in the top of a tall tree curled up, sound asleep, in an abandoned hawk nest.

Each time our fox found a bone or a piece of food it carried it out of the moonlight into the shade of the tree trunk, eating it where it was hidden in shadow. In its ranging course over the snow it not infrequently halted, lifted its pointed nose to sniff the air, peered intently, its ears moving to catch some faint or distant sound. Poised there, it seemed superbly alert, supremely alive. At the first hint of danger it melted away.

So the fox roams through the night, so quick, so aware, so free and independent, so self-reliant, a predator but not a parasite, living by its wits, a freebooter among the wildlife. The fox's character, like its face, is whittled to a point. All its faculties are brought to bear on the problems of survival. Its bright eyes and slender face are filled, if not with a kindly or philosophical expression, with awareness and intelligence. It is through beauty of form and intensity of life that the fox stands out among our visitors from the winter woods.

In our region the gray fox predominates. On rare occasions a red fox, with its white-tipped tail, appears under the apple tree. Half a century ago, it was the red fox that was common, the gray that was rarely seen. Infectious mange spread from hole to hole and almost wiped out the red-furred animals. The gray foxes, which never enter a red fox burrow, escaped the epidemic.

Some evenings a visiting fox will feed on a banquet of odds and ends. I watched one eat a bit of steak fat, some cottage cheese, half a slice of buttered toast, pieces of bran muffins, two small apples, and a final helping of cracked corn put out for the birds. From our bedroom window, late one night, we saw a fox that was feeding on birdseed on the wall under the hickories silhouetted against the starlight. On another night we watched a fox on the wall top in moonlight while a skunk, gleaning seeds scattered on the snow, moved about in the shadow directly below it.

We often observe such amnesty in the winter nights. When I switched on the floodlight at the kitchen door one evening, I saw a skunk and a fox feeding only three feet apart. The fox sat eating with its back to the skunk. Animals appear to sense how other animals feel. There are some nights when everything will give a skunk a wide berth and other nights when little attention is paid to it. A fox, I have noticed, seems most at ease when the skunk is facing it. I have seen one stretch out its neck and seize a scrap of meat within two feet of the nose of a feeding skunk. But when a skunk turns tail-to or raises on its hind legs—cocking its spray gun, so to speak—the fox retreats.

At times when they are not seriously aroused, skunks may make a little rush toward some other animal, a kind of bluffing movement with mouth open and a low accompanying "Woof!" But when they begin to stamp their forefeet and lift their hind quarters it signals a different mood. I recall a large yellow cat that came to the yard for several nights. The first time I saw it in the floodlighting it was crouching in the grass watching intently the movements of a skunk foraging ten or twelve feet away. After a while the skunk became aware of its presence. It peered at it, then made a little woofing rush toward it. The cat pranced away no more than a dozen feet and stopped. The next night I saw the yellow cat back in almost the same spot. On this occasion it was watching a different skunk. When this skunk caught sight of it, it whirled and lifted high its hind quarters. There was no prancing retreat this time. The cat got the message; it understood the language of the skunk. It was still hightailing it across the yard when it passed out of the range of the floodlight.

One evening, in March, about eight o'clock, a lone skunk was nosing about looking for food under the apple tree when a family of raccoons arrived. The three half-grown young were ravenous. Earlier that evening I had thrown out a handful of peanuts. Vying for these prizes, the young animals hurried this way and that, bumping into the skunk, pushing it aside, sweeping it along in their midst as they twisted and circled in their search. The buffeted skunk danced lightly about, lifting high its tail. But it seemed to appreciate the fact that the raccoons posed no personal threat, had no interest in it, were intent only on finding food. So it withheld its fire.

There is, in the gaze of a skunk, something innocent and childlike. Yet there is also, from time to time, a knowing glint. It has a secret. It is not entirely an innocent set down in a wicked world. In its own way it has an ace up its sleeve. But, as a rule, it hesitates to play its trump card; it hesitates to waste its liquid ammunition. That malodorous fluid is not only overpowering in the nostrils, it produces a burning sensation when it comes in contact with the eyes. A dog near here was once seen rushing about in a field, plowing along with its head thrust beneath the snow. A skunk had fired point-blank into its face and its eyes must have seemed on fire. When a neighbor of ours was sprayed in the eyes by a skunk, he thought he had been blinded. But the next day, he said, his eyesight was so clear he "could see all the way to New York." In fact, under certain conditions, the fluid of skunks has been used by oculists for aiding eyesight.

Thinking back over the skunks we have watched at Trail Wood, I remember one that flattened itself out on the ground while it ate, its fur swept this way and that, its plumed tail carried first to one side and then to the other by the rush of the wind on the edge of a hurricane. There was a skunk that came regularly in summer to feed on insects that fell to the ground under the floodlight. We were surprised by the speed with which it slapped down its forepaw on a quarry or, at times, darted about like a kitten catching crickets. I remember two skunks that took to their heels side by side. As though jiggled at the ends of rubber bands, they bounced up and down as they ran. Particularly tame and confiding was another skunk. Being careful that I was always dealing with the same animal, I gradually gained its confidence. In a mildly daring adventure I ended up by being able to feed a wild skunk pieces of doughnut from my hand.

But of all these memories the one most vivid is that of a family of skunks. They came sweeping around the corner of the house on an evening in July. At first glance the close-packed group appeared to be a black and white rug slipping on a wandering course over the grass. The plumy tails, one large, the rest small, waved above the little mass of bodies. Usually young skunks tag along after the mother single file. Here, perhaps on their first expedition away from their home burrow, the young animals, pressing close, followed their mother's every change in direction. The whole family turned and twisted as a unit while the search for food continued.

In watching skunks feeding, we have the impression their search for scraps is a hit-or-miss affair. They wander about, zigzagging like foraging ants. Sometimes they pass within inches of food without detecting it. When they come upon a piece of meat or muffin, they appear to stumble on it by chance. Their seemingly aimless trail, however, may be the most efficient path they could follow. In the end they cover all the ground and nose out whatever food is there.

Among themselves, we have noticed, skunks have a curious method of settling squabbles over food. When one finds a titbit and another tries to take it away, they do not ordinarily growl and bite and scratch like cats or dogs. The second skunk merely moves alongside the one with the food, approaching from the rear. Then they both brace their feet and begin pushing sidewise. They sway back and forth like two wrestlers in this silent test of strength until one is pushed away. The longest contest of the kind that I have seen occurred on a night of gentle rain when I had put out a great prize for skunks, an egg that had been accidentally cracked. The first skunk to come upon it crunched the shell and began eagerly lapping up the contents. It was only partially finished when a second skunk smelled the egg and the pushing match began. Minute after minute it continued. In the end it was the first skunk that was shoved aside. We saw it wander off in search of other food while the victor consumed the rest of the egg.

When dew lay heavy on the grass one evening, and the animals were leaving their small wet pawprints on the flagstones outside the door, we witnessed a variation of the pushing game. Using a piece of ham fat it was eating as a hub, a feeding skunk revolved like a wheel in slow motion, shouldering away a second skunk, blocking it from reaching the prize. In some kind of disagreement, on another night, two skunks, tails up, forefeet patting the ground, danced about in a graceful performance, as in a skunk ballet. In contrast to that two other skunks put on a ludicrous exhibition when they accidentally backed into each other while they were feeding on scattered bits of a muffin. Instead of retreating, both lifted their tails and shoved backward. For a minute or more this tail-to-tail struggle continued. Then one of the skunks, perhaps having eaten its piece of muffin, moved away and the contest was over.

Almost always our skunks consume what they find where they find it. Rarely do they pick up food and carry it to a secluded spot in the manner of a dog or fox. The nearest approach to this that I have seen was when a skunk moved a muffin ten or fifteen feet down the slope under the apple tree. It did this by backing up and rolling the muffin after it with its forepaws.

As food grows scarce toward the end of winter, each year, our visitors from the woods multiply. Gray squirrels, feeling the pinch of hunger, come increasingly to the cracked corn. By winter's end their numbers may reach twenty-five or thirty. One year they rose to nearly fifty. As early as February we begin to notice female squirrels that are partially bare around their shoulders where they have pulled out fur to provide a soft lining for their litters in nests in hollow trees. Some evenings, as the winter is drawing to a close, when I switch on the floodlight I look out on a veritable menagerie. In the course of a few hours I have sometimes seen three foxes, five skunks, two opossums, and four raccoons all within the angle of the yard just outside the kitchen door.

When foxes and raccoons feed together, it is the raccoon that stands highest in the pecking order. The fox is faster but the raccoon is heavier and stronger. Oftentimes with their delicate forepaws—

almost as useful as the forepaws of a monkey—our raccoons will open the mesh suet bags we hang out for woodpeckers. Deftly they will extract the suet from inside without injury to a single strand of the enclosing bag. On other occasions, when hurried or particularly hungry, they may cut off the whole bag and carry it away. Half a mile from here, one winter, a house was left untenanted. When the owners returned in the spring, they found a raccoon had climbed down the chimney, had slept on a pile of soft blankets, and had licked all the labels off the wine bottles in the cellar.

In 1967, after a sixteen-inch fall of snow in March, I dug down to the ground and cleared out a space five or six feet across for putting out food for birds by day and for wild creatures from the woods by night. Soon after dark, that evening, a skunk came up the lane, following the packed-down trail of my snowshoes, and discovered the scraps of meat in this opening. It was feeding placidly when a forlorn opossum appeared toiling across the drifts. It came to the edge of the shoveled-out area, halted, stretched its long pointed nose downward, and sniffed the air. The skunk ignored it. Moving awkwardly along the curve of the rim, the opossum began looking for a way down to the food it smelled. As it blundered along, the snow suddenly gave way and, in a miniature avalanche, it came tumbling down to the bottom. The skunk lifted its tail and raised its hind quarters. But the warning was lost on the opossum.

Struggling to its feet, it peered about through eyes that seemed near-sighted. Then it stretched out its neck until the tip of its nose was no more than an inch from the muzzle of the skunk's liquid gun. Each second I expected the gun to discharge and the opossum to pitch head over heels backward. But nothing happened. The opossum made no quick movements that would have alarmed the skunk. After a long pause, while it seemed to be collecting its wits, it pushed past, almost shoving the skunk out of the way, and reached a piece of meat. Its very dumbness, its slow-witted unhurried movements, probably saved it. At any rate, we saw the two animals feeding almost side by side until all the meat was gone.

An opossum's greatest asset in finding food is its sense of smell.

We often see one thrusting its long nose deep into the soft snow to retrieve a buried bit of bread or meat or doughnut. Usually, as it eats, an opossum holds food in one pink forepaw as in a hand. It never sits up, holding it between both forepaws as does a squirrel, a woodchuck, or a raccoon. Each time one of these animals comes to a piece of buttered bread thrown out on the snow, it follows the same procedure. Beginning at one side, it licks the hardened margarine, carefully rolling it up toward the other side. Only after it has devoured this roll, satisfying its craving for fat, does it begin feeding on the bread.

With their slow movements and their dim-witted appearance, their fur so thin they seem partially bald, their ears and tails exposed to the freezing weather, winter opossums give the impression of animals that have been short-changed in almost everything. As January gives way to February and February to March, they become more wretched-appearing and forlorn. Frequently we see how their ears and the tips of their tails have been severely frostbitten. During the harsher times of winter, the opossum seems unfit to survive for more than a few days. Yet it has extended its range northward in recent times until it is now found in upper New Hampshire and—above Niagara Falls—across the line in Canada. As our only marsupial, its lineage is the most ancient, it is the most primitive, of all our furred wild inhabitants. No other mammal in North America has been on earth so long.

But it is the race of opossums, rather than the individual animals, that carries on. In our region, few survive more than two years. Most of those we see have been born in the year in which we see them. All, however, in spite of hazards and handicaps and injuries, cling tenaciously to life. Late one winter, an opossum arrived in the yard night after night dragging one of its hind legs. It appeared to be broken. For these primitive animals the short span of their lives is passed in an accident-filled world. When opossum skeletons were studied in one museum, it was found that they contained countless fractures—many involving serious injuries—in which the bones had knitted.

Nothing is more dramatic in the constant struggle for survival of these animals than their feigning of death. This is the best-known fact about them. For a hundred years controversy has raged over whether "playing 'possum" is an involuntary cataleptic state resulting from psychological changes in the animal or is merely a sham performance begun and ended at will. Modern scientific aids have, at last, provided a definite answer. In a Los Angeles research center, electrodes implanted in the skulls of fifteen opossums recorded their brain waves before, during, and after this counterfeiting of death. They remained unchanged. The "dead" opossum was merely putting on an act.

From a more simple observation at the other end of Hampton, a few years ago, the same conclusion could be drawn. When one of these animals was caught cutting down a suet bag, it apparently dropped dead. It lay rigidly on its back. Its mouth hung slackly open. Even when it was picked up and carried on a shovel into a field, it lay inert and apparently lifeless. But while it was being transported one of its eyes popped open. It was keeping track of what was going on.

Death-feigning aids the individual. But the paramount secret of survival for the species is rapidity of reproduction. A female opossum begins breeding before she is fully grown. The period of gestation is the shortest known among mammals, only about twelve and a half days. Each litter runs from eight to eighteen with about seven surviving. One of the most amazing stories of zoology is provided by the early life of these baby animals. At birth they are only about one-ten-thousandth the weight of the mother. Twenty will fit easily into a teaspoon. Although they are blind and almost embryolike, they find their way—clinging with strong foreclaws—to the teats in their mother's fur-lined abdominal pouch. Within its protection the baby marsupials remain for more than two months, feeding and growing. When they emerge, they are about the size of mice. For a time, before they are completely on their own, they ride about on their mother's back, often with their tails wrapped around her larger tail which is held arching forward.

In many ways these almost naked prehensile tails come into play during the life of these animals. Looking toward the north wall one afternoon, I saw an opossum moving slowly along the top. It reached the barway and began to walk deliberately across the topmost pole. At the halfway point it stopped, wrapped its tail around the pole, and let itself down headfirst to the pole below. In trees and among vines, the opossum's tail provides a safety anchor checking its fall if it loses its footing. And on several occasions scientists have observed these animals in a remarkable employment of their tails when they have been gathering leaves for bedding. Picking up mouthfuls, they push them back between their forelegs onto the tail which is curled under the body. When half a dozen such mouthfuls have been packed into place, the tail is wrapped tightly around them and in this way they are transported back to the nest.

I suppose to speak of a beautiful opossum seems a contradiction in terms. But one fall we were visited by a strikingly different darker-hued, partly melanistic individual that came nightly for food. Its head was white, its body dark, almost black, its nose and feet pink. In the floodlight its three-toned coloring stood out, strikingly handsome, among the other animals.

The most beautiful of all our woodland visitors? It appeared in falling snow one late-November afternoon. From the woods beyond the waterfall, it had come down the path beside the brook and stood beneath the apple tree at the foot of the slope toward the east—a magnificent antlered buck with head held high, its body smooth and muscled, at the peak of its autumn health. Surrounded by the drifting descent of soft goosefeather flakes, it remained motionless. It was no more than fifty yards from our windows.

As we followed its movements with our eyes, it nibbled tentatively at the dry and yellowed leaves of a pokeberry. Then it turned toward us and wandered up the winding paths of my Insect Garden. We saw it come to the breast-high stone wall that forms the boundary of the lane. It paused for only a moment. Then, with no apparent effort, it soared in an upward arc over the barrier and into the lane. This it crossed without haste to the farther wall. Once more the thrill of its almost drifting leap and it landed in Firefly Meadow. So graceful, so unstrained was each of these effortless jumps that the great animal seemed riding the rise and fall of a draft of air. We followed its form as it moved away across the meadow, followed it until it grew gray, dim, then faded from sight entirely amid the falling snow. Long after it had gone—even today—the memory that remains seems symbolic of all the beauty and wildness of our woods.

Some years we see no deer at all. At other times we have seen as many as five at once beside the pond. Along our trails we often notice where they have nipped the tops from blackberry canes or have browsed along the sides of the red cedars. In the James L. Goodwin State Park, on the west side of Hampton, one spring, a young buck was observed feeding on pussywillow catkins. It would wrap its tongue around the base of a branch and then strip off all the aments at once with a sidewise sweep of its head. Under the wild apple trees in the midwinter woods, we sometimes find where deer have dug down through the snow to reach the dried fruit that fell in autumn. And once, at the end of winter, we discovered, under Old Cabin Hill, that the Lost Spring had been cleaned out by the pawing forefeet of the deer.

Like the foxes, like the skunks, like the raccoons, like all the wild creatures that visit us on winter nights, the deer connect us with the earliest days of this farm. Their kind was here at the beginning—and long before. The pegged beams and the old fireplaces are inanimate links with the past. But the animals are living links with all those other times.

CHAPTER TWENTY-ONE

Firelight Nights

Where the terraced kame of the North Cemetery nears the foot of its steep descent toward the south, a square white column of weathered stone marks a grave that is more than a century old. Here lies the beautiful first wife of Andrew J. Rindge. And here, nearly fifty years later, the poet was buried beside her. The more recent inscription is clear cut on the column's southern side: "Andrew J. Rindge 1835–1912." On its eastern side, now blurred by time, is the earlier legend:

GALISTA S. F.
WIFE OF ANDREW J. RINDGE
DIED OCTOBER 29, 1866
AGED 27 YEARS
4 MONTHS AND
8 DAYS.

Below, nearly erased by the storms of a century, appear the poet's simple, moving lines:

> *A light is from our household gone,*
> *A voice we loved is stilled,*
> *A place is vacant at our hearth*
> *Which never can be filled.*

The hearth of which he speaks is the one beside which we sit on winter evenings at Trail Wood. It has become part of our lives as, so long ago, it had become part of the life of the ill-starred poet.

Before there was a Trail Wood house, there was a Trail Wood fireplace. Before there were any walls or a roof, there was this massive central chimney. It was the custom in earlier times to complete the chimney first and then build the house around it. As a regional saying has it: "A chimney isn't a chimney unless you can walk around it." When it is placed at one end of a house, a chimney dissipates much of its warmth outdoors; incorporated within the structure, it employs virtually all its heat in warming the rooms around it. Thus a central location not only permits it to carry away the smoke of more than one fireplace but it results in the most efficient use of the heat.

Built in the year the Lewis and Clark Expedition was returning from the mouth of the Columbia River, our chimney serves four fireplaces: the large one in the living room, nearly five feet long and four feet high; two medium-sized fireplaces, one in the south bedroom and the other in my book-lined study; and a miniature fireplace—designed not for burning wood but for holding live coals—in the second-story bedroom at the head of the stairs. The floor of this room, formed of wide boards held in place with square hand-wrought nails, has all its cracks filled with clay. This precaution, dating from Colonial times, prevented live coals from dropping into the open spaces between the boards. Where the chimney rises through the attic above the fireplaces, the bricks for more than a century and a half have been held together with clay instead of mortar. During the 168 winters since they were set in place, the smoke of perhaps 10,000 fires has ascended past them.

For us the pleasures of our fireplace begin even before we light the first fire. During the latter days of October and the early days of November, Nellie and I range through the woods over fallen leaves, gathering sticks, breaking up dry branches, picking up poles in a harvest of winter kindling. It is as little work and as much fun as going nutting in the fall. We carry home our loads across pastures filled with the sound of crickets and often with bluebirds calling overhead. At times we follow the path over the Starfield balancing long poles on our shoulders, making a game of it, seeing how far we can walk without touching the poles with our hands. We find something deeply satisfying about this annual gleaning in the woods. We seem obeying some primitive instinct to store up for the cold months ahead.

Then, in the short winter days and the long winter evenings, the great fireplace of our living room comes into its own. It brings light and color and movement and sound and perfume and a direct warmth into the room where an old wall clock ticks away the minutes and chimes the hours and half-hours throughout the day and night. At first, before we had it cleaned and checked, its double-strike had a rhythm that brought to mind the calling of a whippoorwill. Once when a visitor was afraid the unfamiliar striking of the clock might keep him awake during the night, I stopped the pendulum when he went to bed. The result was that, in the unaccustomed silence, Nellie and I lay awake much of the night.

The appeal of an open fireplace is deep-seated. It has its roots in four of our five senses: sight, hearing, feeling, smelling. We watch the flicker and the altering shapes and colors of the flames. We hear the snapping and crackling of the burning logs. We smell the perfume of the various woods as they are consumed. We feel the warmth of the dancing flames and glowing coals. Endlessly these elements are combined and recombined. No two fireplace fires are ever the same. Each represents a different pattern of flames, a different sequence of sounds, a different play of colors. These fires of winter are as dissimilar as wave marks on the seashore, as varied as autumn leaves or flakes of snow or human beings.

The voices of our fireplace range from a soft flutter of flames, like a silken flag flapping in the breeze, through sharp snappings of the burning wood, like small firecrackers exploding. At times there are tiny cracklings like sleet on a windowpane. Then, at the end of the evening, come the sleepy-sounding fires, dying, falling into silence with a soft simmer and murmur as lulling as rippling water or rustling leaves.

How wonderfully snug and enclosed we feel in winter storms with logs blazing on the hearth! Sitting there, gazing at the ever-changing kaleidoscope of the flames as they flicker before the smoke-blackened stones, we often become aware of a curious dislocation in time. We might be enjoying this warmth and light and color in any other period during the long history of this companionable hearth—before the first airplane flew or before the Civil War or when California was in Spanish hands. In no other hours is this feeling of temporarily being afloat in time, of living in undated moments, more apparent than when Nellie, filling in gaps in our acquaintance with the classics, reads aloud at the end of the day from books that came into being over a span of centuries of time.

And as the winter days go by, as the logs of oak and maple and hickory that are packed row on row in the center shed, each in turn burn to ashes, we watch the piles dwindle down like sand in an hourglass. Each year we burn about five cords of wood in our living-room fireplace. The gradual disappearance of our fireplace wood measures, as in some larger glass, the progress of the season. In all varieties of winter weather I bring in the logs. Often as I emerge from the shed, a log on my shoulder, the smell of woodsmoke is sweet in the clear cold air.

Each in its own way, the wood I carry in burns on the hearth. Part of our continuing enjoyment of these firelight nights is discovering the special qualities of different fuels. A limb dies or a tree falls and we experiment with its wood in the fireplace. For a bright steady colorful flame and long-lasting coals our favorite is apple wood. It is our choice for broiling steaks and baking potatoes. White oak and sugar maple and well-seasoned hickory also burn with a steady flame. In heat production, a cord of pignut hickory is equivalent to a ton of high-grade anthracite coal. Willow, which is almost never used in fireplaces, is slow and sluggish in burning. It is this characteristic that makes it particularly suited for the production of charcoal. Ash reverses the usual rule. It burns as well when it is green as when it is seasoned and dry. The explanation is that it is related to the olive tree and its wood, when green, contains one of the constituents of olive oil, oleic acid. Pine logs also burn well when green for the resin they contain produces a hotter flame than the wood itself. But, like the wood of green ash and all the resinous evergreens, such logs give off a gummy smoke that rapidly builds up soot in a fireplace chimney.

In some instances the color of the flame is the special feature of the wood we burn. Oftentimes we turn out the lights to enjoy it better. White birch, for instance, blazes up with a clear and brilliant yellow flame. On occasion friends have brought us bits of driftwood impregnated with the minerals of the sea. When we add them to a fireplace fire their flames weave in a flickering rainbow display of metallic blues and reds and greens. Other woods we add to the fire in order to enjoy the perfumes they produce: chips of mesquite brought back from Texas, small poles of black birch to add their sweet smell to the room, and always twigs and branches of the hickory trees.

"What shall we put on next?" is a frequent question of these winter nights.

Green hickory, as it is consumed, produces the most music of any of our native woods. It sings, as John Burroughs once said, "like winter wrens." But to produce the effect of a chain of firecrackers, toss on a dry stick of butternut. One time when our publisher-friend Raymond T. Bond visited us, he brought along a classical record as a present. We had just put it on the player when Nellie added a stick of wood to the fire. As soon as it began to burn, she realized it was butternut. We heard the record only intermittently through an almost continuous fusillade of explosions. Also noted for giving off sharp pops as it burns is American chestnut. Nearly half a century after one of these great trees died in the blight that swept through our woods in the 1920s, I put a fragment of a weathered stub on the fire. It still popped loudly as it burned.

An old Central European superstition concerns the sparks that rise from a fireplace on Christmas Eve. Blessings as numerous as the sparks are supposed to descend on the family during the ensuing year. A believer in that superstition would find special comfort in using mulberry wood. Not only does it burn noisily but its long series of explosions are attended by successive showers of sparks that go streaming up the chimney.

All during one Thanksgiving Day at Trail Wood the fire that burned on our hearth consumed a single log, the last immense section of a long-dead elm tree that had been polished by wind and weather to a silvery hue. At the end of the day all that remained was a ghostly pile of fine white ash. In this almost weightless drift of residue, the history of a great tree had come to an end. Sitting there, watching the surface of the ashes stirred by each small current of passing air, I found myself trying to recall the exact words of something I had read long before. Later I looked it up. It proved to be that sonorous passage from a seventeenth-century sermon of John Donne's that begins: "The ashes of an oak in the chimney are no epitaph of that oak, to tell me how high or how large it was: it tells me not what flocks it sheltered while it stood; nor what men it hurt when it fell."

The fastest-burning material I have ever thrown into our fireplace was a handful of the infinitesimally small spores of the club moss, *Lycopodium complanatum*. When I tossed this living yellow dust, gathered in the woods, among the flames, it ignited in an almost instantaneous flash of brilliant light. This same light, produced in a similar way, stabbed across the stage in Shakespeare's day when lycopodium powder was ignited to simulate lightning.

On an evening in autumn, when one of the earliest fires of the year was burning on the hearth, a small tan-colored forest moth found its way into the house. We watched it flutter about the living room and sweep down the length of the massive fireplace. Most of the stones there are gray, formed of Canterbury gneiss. But among them, near the right-hand side of the fireplace, there is a single block of stone tinged with tannish-yellow. The flying insect passed across the gray rocks until it came to this yellow-tinted stone. Immediately it landed. It seemed to disappear, merging with its background as soon as motion ceased. Nellie and I looked at each other in amazement. How had it picked that particular stone? How did it know what color would camouflage it best? How did it recognize that, among all those blocks of gray gneiss, this one stone provided the place where it would be least conspicuous?

The explanation that I have heard is that its ability lies in the character of its eyes. They bulge out like round mounds made up of many lenses. Those at the rear overlook both the leading edges of its wings and the background passing beneath them. When the moth is in search of a landing place, if the colors are different it keeps on flying. But when they match or are closely similar, an automatic response is triggered and the insect alights.

This hypothesis is so simple, so logical, so easy to understand that I was almost sorry to learn that recent tests have cast doubt upon it. At the University of Massachusetts, Dr. Theodore D. Sargent asked himself: What would be the effect if a moth's wings were painted over, changing their hue? Using light-colored and dark-colored *Catocala* moths, he discovered that when he painted the forewings of the light moths dark and those of the dark moths light, the insects still chose resting places that matched their original colors. Thus their capability in selecting areas where they will be best camouflaged remains a mystery that defies easy solution. It lies in that mysterious realm of natural abilities that are, as scientists say, "genetically fixed."

In late September and early October each year, as the weather turns chill, our fireplace becomes a vertical highway for houseflies. Especially before a general storm, the insects seem suddenly everywhere in the house. Mosquitoes, at times, also find their way down the darkened flues. But of all the winged creatures associated with the natural history of an old country fireplace, the chimney swifts are the ones that attract the most attention.

They appear early in May and are gone by mid-September. Between the time we see them one year and the next, they travel to another continent to spend their winter amid the jungles of the upper Amazon. Their crackling high-pitched calling heralds their return in spring as they whirl about the house and parachute down on set wings into the open mouth of the chimney. We hear a diminutive airy thunder as, within their narrow confines, they brake their descent with fluttering wings. This is a sound that repeats itself a thousand times during the weeks of summer. We even hear it long after midnight at times when the moon is full.

It is joined, as the season advances, by another sound, the shrill buzzy chittering of the nestling swifts. As the young birds grow older, their calling becomes almost continuous. It forms overlapping waves of sound that fill the dark passageways of the chimney, chatter through the rooms below, and carry for a surprising distance out-of-doors. On still days, I can sometimes hear the calling of the young birds when I am sitting under the apple tree at the far side of the yard. It increases and fades away, then rises to a fresh crescendo as a parent bird arrives with insect food.

In our second year at Trail Wood, on the last day of June, an afternoon of sultry heat ended in a violent thunderstorm and a deluge of rain. When it was over, I found that a baby swift, in pinfeathers, had fallen from the nest. Even at its early age, an immature swift possesses a remarkable ability for climbing. I recovered the young bird and held it as high as I could in the chimney and let it anchor itself to the masonry. We never saw it again. Apparently it succeeded in hitching itself gradually upward toward the nest.

In connection with young swifts, an odd discovery was made by accident, a few years ago, on Long Island. An ornithologist, attempting to feed a brood that had fallen with a dislodged nest, found that nothing he could do would make the little birds open their bills. Then a breeze, blowing in through an open window, struck them. Instantly all their mouths popped open. The sudden rush of air within the chimney, caused by the braking flutter of the parent birds when they arrived at the nest, appears to be the "releaser"—the signal for the nestlings to open their mouths and be fed. After that, all the ornithologist had to do to make them open their bills was to blow a breath of air over them.

Toward the end of summer, our swifts circle the house on wings that are frayed at the ends from many comings and goings within the flues. Like hawks, swifts depend on their wings for food. And both birds shed their flying feathers two at a time, one from each wing. When migration begins, our swifts are clad in new plumage. They are at their aerial best. For days before that epic event, the sound of wings and voices increases in our chimney. Within those little heads, so few yards from us, there is a growing restlessness—a hereditary pull like the tug of the sea for the eel, of the shore for the spawning horseshoe crab. Then suddenly the swifts, old and new, are gone. The baby birds we heard chittering on their nest a few weeks before are on their way to the Amazon. Our chimney is left vacant and still.

Not long after this, we light our first fire in the fireplace. All through the succeeding weeks of winter the flutter of the flames, the snap of the burning logs, the warmth of the fire are features of our living room. During the time marked by these successive fires, the days gradually lengthen. At last the final log comes from the shed, the last fire burns itself out, the last ashes are cleared from the hearth. Winter and the fires of winter are behind us. Firelight nights are over for another year. Once more the chimney is left to the swifts.

CHAPTER TWENTY-TWO

Birds of the Winter Wind

Three hundred wild birds—redpolls, juncoes, pine siskins, white-throated sparrows, evening grosbeaks, purple finches, tree sparrows, chickadees, goldfinches, tufted tit-mice, mourning doves, bluejays, cardinals—are feeding around our house on this cold and windy morning in February. Thus day after day, in snow and ice, in sunshine and under heavy skies, while the thermometer rises and drops, all through the winter, cold-weather birds, the birds of the north wind, surround us.

We note their varied ways of feeding—the jays stuffing their throats with as many as twenty sunflower seeds before flying away; the mourning doves picking up scattered grain, their heads moving so rapidly up and down they bring to mind sewing machines in action; the evening grosbeaks snatching up sunflower seeds from a feeder, turning to face outward, running the seeds sidewise through their large ivory-yellow bills and letting the hulls and fragments fall to the ground. There the little striped pine siskins run about like feathered mice, feeding on the crumbs from the grosbeak's table. In January storms, when drifting snow has covered all the scattered seed, we sometimes see juncoes clinging like chickadees to the rough bark of a tree trunk to pick out bits of suet that we have rubbed into the crevices.

One source of amusement on these winter days is watching the bluejays trying to get mouthfuls of the suet we put out for the woodpeckers in hanging mesh bags that dangle below the limbs of the apple tree. We see them learning by experience. One jay at first pecked at the swinging bag as it was moving away from it. Then it discovered it got in a harder peck and a larger mouthful of food if it waited until the bag was moving toward it. Other jays learned to cling upside down to the bag, getting several mouthfuls before they let go. One of these birds had difficulty maintaining its position. We saw it waving one wing to balance itself. This wound up the bag so that, when the bird flew away, it unwound like a spinning top.

The intelligence of jays in solving problems connected with feeders is proverbial. I think the most dramatic instance of the kind of which I have heard occurred near Fairhaven Bay, in Concord, Massachusetts. A friend of ours, Esther Anderson, installed a feeder in which a doughnut could be pecked at by small birds alighting on a circular wire perch. If a larger, heavier bird, such as a bluejay, alighted on the perch, its weight moved the doughnut out of reach within the protection of a cuplike shield of plastic. After being baffled over and over, one jay evolved a way of outwitting the mechanism. Clinging to the perch with one foot, it beat one of its wings rapidly. This lifted most of its weight from the wire and the doughnut reappeared within reach of its bill.

Whenever we open the door or step outside these winter days, we hear the voices of the birds. We listen to the loud trilling calls of the grosbeaks, the tinkling little bell songs of the juncoes, and the ascending "Shreees" of the pine siskins. We catch the bright "Perchicory" of the goldfinches, sometimes coming down through falling snow; the drawling, whistled "Phoebee" of the chickadees during winter thaws; the sweet jingle of the tree sparrows at sunset like chimes of tinkling icicles. We notice how the juncoes and the tree sparrows are stimulated to sing just before a snowstorm. During these days in the dead of winter, those lines from Oliver Herford's poem about the bird singing in "the dark of December" often come to mind. For us, too, the voice of the bird among these bleak surroundings is "a magical thing and sweet to remember." In truth it reminds us that "we are nearer to spring than we were in September."

Looking out on the coldest days, we watch the birds sitting on their feet, fluffed up against the cold, blown about by the gusts. We see them picking at dirt turned up by a snowplow along the lane, seeking essential gravel for their gizzards. We observe them digging in the snow for food, the jays and mourning doves with sidewise sweeps of their bills, the tree sparrows and juncoes scratching with both feet at once. One junco kept turning like the hands of a clock as it scratched, sinking gradually deeper and deeper into its excavated bowl or cup in the snow. We were amused to observe a cardinal hopping about close by, snapping up seeds thrown out by the smaller bird. Once a bluejay struck a vein of buried seed in a snowbank and followed it farther and farther until we saw the bird disappear entirely within its self-made cave. Occasionally mourning doves will excavate a hole so deep only the tips of their sharp-pointed tails are visible rising above the edge.

One hairy woodpecker comes flying from the woods when it hears us pounding suet into the crotch of a tree. Hammering is part of the language of a woodpecker and to it this particular sound speaks of food. It is like a dinner bell to the suet-eater. Many of the small ground birds, we notice, particularly the juncoes and the tree sparrows, choose sides, feeding regularly either on the east or the west side of the house. Sometimes on cold mornings, after I have first scattered seed on the east side and the birds are feeding there, I find half a hundred others waiting on the west side, still unfed but remaining where they are accustomed to find their food.

From year to year and from day to day, the number of individuals and of species rises and falls. In the bitterest times of winter, after ice storms and blizzards, we sometimes find we are feeding more than

400 birds. The figure for the mourning doves has risen as high as 184; for the evening grosbeaks as high as 118; for the redpolls, in years of abundance, as high as sixty-four. These counts were made of birds gathered in the yard at the same time. How many individual birds visit us in the course of a winter season we have no way of knowing.

We count our chickadees over and over. But how do we know we are counting the same ones? They are not banded; they are as alike as the proverbial peas in a pod. When Dr. James A. Slater, of the University of Connecticut, banded chickadees in his yard a few years ago, he thought he had about a dozen of these birds visiting his feeders. Actually, his banding revealed, the number was more than eighty. Different chickadees were coming and going throughout the day. Similarly when another University of Connecticut ornithologist, Dr. Jerauld A. Manter, banded pine siskins one winter, the highest number he counted in his yard at any one time was about twenty. Yet the number he banded there exceeded 100.

Many winter birds, grosbeaks and finches particularly, are given to wandering. However such movements usually go unnoticed unless some special marking or deformity makes the individual easy to recognize. One December an evening grosbeak that had lost one leg appeared at Trail Wood. During succeeding days it returned to our sunflower-seed feeders. But during those same days it attracted attention at other feeders, one two miles away, another four miles away.

Have you ever seen a chickadee without a tail? We have. But only once in our lives. It appeared on a winter day outside our kitchen window. Without any apparent difficulty in flying, it darted like an overgrown bumblebee from tree to feeder and back again. In split-second escapes from enemies, birds not infrequently leave their tail-feathers behind. We have seen tailless catbirds and mourning doves and towhees and brown thrashers. At the Moosehorn National Wildlife Refuge, on the upper coast of Maine, some years ago, an occurrence demonstrated that in emergencies birds can voluntarily shed this part of their plumage. A ruffed grouse was accidentally caught in a live trap set for woodcock. When a biologist reached inside to release it, the bird thought it was caught by an enemy. As he lifted it out, the scientist saw its whole tail fall off "as if attached to a string that had been cut."

Before the tail feathers of our chickadee grew out again, the little bird had added its bit to the evidence of wandering in winter birds. For during the same days we observed this tailless mite, a chickadee lacking a tail also appeared at the feeder beside the library in the village, two miles farther south along the ridge. Was it the same bird? The chances seem overwhelmingly in favor of that assumption. Inasmuch as we have never seen another tailless chickadee in a decade and a half at Trail Wood, or in our whole lives elsewhere, the coincidence of two chickadees losing their tails at the same time, in the same locality, appears too great to be accepted.

Toward the end of another winter, when food had grown scarce, we saw a red-shouldered hawk sweep down across the yard and alight on a piece of suet in our apple tree only thirty feet from the house. Carrying it gripped in its talons, it dragged the suet to the ground. There the hawk covered it with its outspread wings as it would have done with a living prey. For more than a minute, the bird stared fiercely around. Then it fell to tearing out large chunks of the food. With each mouthful, it shook its head violently, sending small fragments flying. At last, with a remnant of the suet shining white in one of its talons, it lifted into the air and headed in a swift straight line for the protection of the North Woods.

Whenever a snowstorm begins, we notice all our smaller birds are on edge. For it is then most often that a sharp-shinned or a Cooper's hawk slips like a speeding shadow out of the gray curtain of the falling snow. The scream of a jay usually announces the arrival of one of these birdhawks. But often the warning comes too late. At such times a scattering of dark little feathers may identify the victim as a junco.

Watching birds come back to their feeding after being frightened by a hawk, we observe different birds have different "return times." The first to reappear are the lively little chickadees and tufted titmice. Then come the tree sparrows and juncoes. Then the evening grosbeaks. It is always amazing to observe the quick recovery of courage in the smaller songbirds. Once I saw a bluejay with a junco on its back in the snow. The jay was hammering the smaller bird with its bill; feathers were flying; the junco undoubtedly would have been killed if I had not flung open the door and frightened away the bluejay. Both birds flew off, the junco in sorry condition, with only two of its tail feathers remaining. Yet, five minutes later, it was back feeding in the same place with bluejays feeding around it.

For three years in succession, we were visited in the heart of winter by what was probably the same beautiful blue-gray hawk. Larger than a crow, the largest of the accipiters, it was a goshawk coming down from the northern woods. Each time it remained about three weeks. Its favorite perch was in the top of the highest hickory across the lane. Birds know their hawks. This was demonstrated dramatically by the reaction of the bluejays. For now there was no bragging, no taunting chorus of screams such as greet smaller and slower hawks. Silently the jays melted away or perched motionless deep among the twisted branches of the apple tree. Once the goshawk streaked down, making a pass at a gray squirrel leaping among the upper limbs. Not a jay flew. But they all, in a wave of movement, shifted their position slightly. So far as we could determine the goshawk caught only one jay. It picked it from the air and returned to its perch in the hickory tree, letting the feathers fall to the ground as it fed. For more than an hour after the hawk had left, the tree swarmed with bluejays, screaming, flying from limb to limb, peering down at the scattered feathers below.

In a few days after the goshawk first appeared, the number of jays at Trail Wood dropped by half. And of the more than 160 mourning doves that had been feeding daily in our driveway, only four—probably resident birds that nested here—remained. All the rest came back no more that year. The danger was too great around our ample food.

While the goshawk will occasionally take a small bird, it is essentially a hunter of larger game. This the small birds appeared to sense. They were far less uneasy than the jays. Day after day, juncoes and tree sparrows fed along the wall just below the hickory where the gray predator from the north was perching. An exception appeared to be the house sparrows. When the goshawk appeared, they collected in a forsythia bush not far from the hickory and set up a peculiar rhythmic chirping, a kind of distraction chorus similar to the one we once heard in the presence of a different hawk at Thomas Jefferson's home in Virginia when we were coming north with the spring.

One February morning I was sitting beside my study window where I am writing these words when a covey of fifteen quail came along the wall under the hickories. I heard their soft "Toy! Toy!" calls rise in a sudden clamor and, looking out, saw them all scaling away down the slope of Firefly Meadow. The goshawk had come back to its favorite perch. It shot down in a strike but it rocketed up again with no prey in its talons. It seemed impossible that all the quail would escape. But on the southward-facing hillside the snow had melted and when they dropped down and froze into immobility among the waves of winter-gray grass, they vanished completely. I swept the slope with my field glasses and could not locate a single one. Yet half an hour later, after the hawk was gone, when Nellie walked down to the frozen pond, quail flew up from all across the hillside.

These little game birds, plump and alert, we see walking up the lane, giving their rallying call from the top of stone walls, picking at gravel where the ground is bare, running belly-deep through the soft snow, sunning themselves at the edge of the plum tangle, tails together at the center and heads radiating out toward all points of the compass. In many of their ways they have an appealing childlike quality. John Burroughs touched on this in a letter to his son, written in the winter of 1900: "I saw the trail of the quails, poor little things, six of them now. They crossed from Brookman's swamp over to ours. Where the woods were dark they drew together like scared schoolchildren; I almost fancy they took hold of hands—real babes in the woods."

Beneath the hickory trees, one midday in December, thirteen quail were lined up in a row, sunning themselves. As I watched them a hen ring-necked pheasant appeared, making its way over the icy crust of the snow, slipping now and then. The last survivor of several pheasants turned loose in the fall for hunters to shoot, this lonely social bird seemed attracted by the flock of smaller birds. At any rate, it attached itself to the covey and remained with the quail the rest of the afternoon.

The life of the pheasants we see usually extends for only a few weeks into winter. These birds, after being cared for like poultry at state game farms, have been suddenly released into a hostile and menacing world. Most are soon caught by foxes or shot by hunters. In late fall, we see the survivors coming to the wild plum tangle for cracked corn, scratching like hens in a barnyard and then, after they have fed, racing back and forth, swerving, turning like whirligig beetles amid the maze of interlacing plum trees. The lone hen we noted joining the quail used to race and twist, stop and turn, for minutes at a time when it had finished feeding. They all seemed to derive some deep-seated pleasure from exercising their inherited ability for running and dodging.

One year, over a period of several weeks, three hen pheasants came regularly to the plum tangle. One appeared to be an outcast. The other two continually tried to leave it behind. They would work gradually away and then race toward the woods. It would awake suddenly to this desertion and set out in hot pursuit. Several times we followed the tracks of our pheasants in the snow and found they most often led from the plum tangle to the brook and along it into swampy, lowland woods. There, even early in the winter, the rolled-up leaves of the skunk cabbage thrust their spearheads above the frozen ground. They provide one of the natural foods for pheasants that survive through the winter in the Northeast.

In contrast to these artificially introduced birds, the ruffed grouse of our winter woods are superbly equipped for survival. They weather the fiercest blizzards by diving deeply beneath the insulating blanket of the soft snow. They walk over the drifts on broadened winter feet, as on snowshoes, leaving behind them tracks that resemble the silhouettes of airborne geese, the mark of the long central toe representing the outstretched neck. And wherever wild apple trees grow in the woods, they find ample fare by browsing on the winter buds. During the latter days of the season I sometimes watch them from my study window feeding in the top of the old apple tree by Veery Lane.

During their harvesting of the buds I see these large brownish birds balancing themselves among the upper branches, tightroping along the smaller limbs, working their way for a surprising distance out into the maze of the farthest twigs.

One evening after sunset, as I brought fireplace logs from the middle shed, I stopped for a moment in the shelter of the lilacs. The wind was rising. The cold of the winter night was pushing the mercury toward zero. Where were the chickadees now? And the juncoes and the mourning doves, the grosbeaks and the tree sparrows? How were they meeting the night of the birds? Each using what protection it could find, each inside some hollow tree or clamped to a twig or limb or other support, each with feathers fluffed for greater insulation, each with its higher body temperature warming it in the dark, all were prepared by form and habits to survive the wind and the cold of the night and to meet whatever the dawn would bring. How hearty and well and enduring seem the lively birds of the winter wind!

But always there are a few unfortunates among them, the sick, the injured, the parasitized. I recall a tree sparrow that was unable to use its right leg and swung its tail far to one side to brace itself and keep its balance on the ground; a robin with a damaged wing that lived most of one winter on Juniper Hill, feeding mainly on juniper and sumac berries; a mourning dove with an injured leg that jerked upward so it goose-stepped when it walked. Even the slightest injury takes on added importance in winter. The saddest of all the unfortunate birds we see are the bluejays wasting away in a lingering death as a result of tongue worms.

These tiny threadlike capillary worms produce nets or diaphragms—something like the effect of diphtheria—in the throats of the birds. They swallow with difficulty. They gape often. They sit fluffed up and huddled in protected places. They remain in the fading light of the short winter days long after the other birds have gone. Dr. Lawrence R. Penner, of the biology department of the University of Connecticut, who has been studying these parasites, reports that death for the birds often comes with torturing slowness. For almost two years, a jay dying of tongue worms lingered on in his laboratory. Almost up to the hour of their death, we have noticed, such birds are able to fly. One of our parasitized jays died in the late afternoon leaning against the bottom of the garage door. Another, as I have mentioned before, came to the end of its life where it had pushed itself into the protection of a small opening in a stone wall. Another, the most fortunate of the three, met a quick death when it was taken by a hawk.

For the healthy bird the chief problem of winter is less the cold than the food supply. With ample nourishment, birds that normally migrate south can overwinter here. For two years in succession a towhee stayed with us. At other times we have had a male redwing, a brown thrasher, bluebirds, and always the goldfinches. An overwintering meadowlark, one year, came to the carriage stone to feed on cracked corn. And, about noon on a February day, a flock of thirty robins alighted across the Starfield to the north. Perhaps, in the mild weather of a thaw, they had come from the coastal swamps where these birds often spend the winter.

Each year it is late October or early November before we begin scattering seed and filling the suet and sunflower-seed feeders. Begun too early, feeding will hold back birds that would otherwise migrate. Always artificial feeding attracts more winter birds than the natural food supply of the area will support. So, once begun, feeding must continue until spring. Few more cruel things can be done to wild birds than making them dependent and then deserting them in the midst of winter. In other words, if you plan to go to Florida, don't begin feeding the birds at all. Moreover it is best not to continue feeding too far into the spring. It holds an abnormal number of birds and numerous species dislike to nest where many other birds are present. We quit feeding, each year, at least by mid-April.

The first thing in the morning and the last thing in the afternoon food is especially needed by winter birds. If possible some of the seed should be scattered in weed patches. There the small birds can obtain it without competing with bluejays and other larger birds. In the midst of our feeding area we always heap a pile of brush or set up an old Christmas tree. This provides a quick refuge for small birds when a hawk appears. Where heavy feeding occurs the ground is sometimes covered with what appears to be abundant food when, in reality, it consists only of the hulls of seeds left by the birds. Misled, people have occasionally declared they would put out no more food until the "lazy" birds finished eating what they had been given—when, in fact, they had already eaten it.

After a storm it is often better to pack down the snow than to shovel it out down to the old feeding level. This gives the birds a fresh, clean table and helps reduce the danger of epidemic diseases. But in one or two places it is well to have bare ground open as a source of gravel. After one blizzard I observed a white-breasted nuthatch obtaining emergency gravel for its little gizzard by picking tiny granules from the composition shingles on our kitchen roof.

Entirely different stomachs are required for eating hard seeds and soft insects. The seed-eaters employ muscular grinding gizzards, the meat-consumers glandular stomachs that digest food almost entirely by enzymes. In the case of a number of birds, and the red-winged blackbird is one of them, seasonal changes occur in the stomachs. The shift from soft summer insects to hard winter seeds is accompanied by a rapid increase in the muscles of the gizzard. The hardness of the food these muscles, aided by bits of gravel, have to grind up within the body of the bird can be appreciated by anyone who ever has had a tiny round millet seed get inside a shoe.

Even when ample food is available, we notice our birds varying their diet in different ways. Downy woodpeckers leave our suet to chisel open goldenrod galls to feed on the hibernating insects within. House sparrows consume the buds of forsythia bushes. Goldfinches turn to nettle and catnip seeds. Juncoes also consider catnip seeds a special delicacy. We see them outside the kitchen window, beside the back steps, leaping up from the snow to flutter in the air and snatch the seeds.

Always after winter gales, across the Starfield and down the slope of Firefly Meadow, we see small birds scattered over the snow, hopping about, picking up seeds the gusts have shaken from the dry rough-fruited cinquefoil. Many plants of the open drop their seeds in installments rather than all at once. Thus they provide food for birds after a succession of winter storms. Each time I throw out grain in gusty weather, I am interested in how the tree sparrows and juncoes go far downwind and feed first on the thinly scattered seeds. They are accustomed, in natural conditions, to finding seeds a few at a time.

To take the pressure from the more expensive birdseed, we put out piles of cracked corn. They attract the hungry bluejays, mourning doves, and house sparrows. They also attract, in the night, cottontail rabbits. At Trail Wood, each winter, we have corn-fed rabbits. We also have corn-fed gray squirrels. Once when we turned on the floodlight at night we saw a large black-and-white cat licking up the particles of ground corn. Dining on this bird food at various times have been opossums, raccoons, skunks, foxes, dogs, and even, on one occasion, a stray horse that had broken out of its pasture.

But of all the creatures that have been attracted to the ground-up corn, the most unexpected were the honeybees. Before the end of winter, in their earliest days abroad, when skunk cabbage flowers and pussywillow catkins were the main source of pollen, we found them crawling over our little piles of corn in search of a pollen substitute. On March 4, one year, I found them all over the yard wherever the corn was present. On one small mound of grain, less than six inches in diameter, I counted thirty-five honeybees. Continually they moved back and forth, gathering corn dust, the pale-yellow vegetable powder which they packed into the pollen baskets on their hind legs.

The main consumers of our corn are the mourning doves. They appear as soon as it is light in the morning, return about noon, and feed for a last time toward dusk—in other words, they have a breakfast, a lunch, and a supper. Each time they concentrate their feeding into ten or fifteen minutes. Then they are gone. Their numbers decrease when thaws open up the fields; multiply rapidly when gales, snow, and ice bring the hardest days of winter. At times they carpet the snow so densely that late-arriving birds have difficulty finding a place to land. Shut in by a storm, on a December day, we watched one circle over the mass of feeding birds, then flutter down. Coming to rest on the back of a dove, it rose with a whistle of wings, then descended again—once more on the back of a dove. A third try and it dropped into a small space among the close-packed bodies.

At times when we are being visited daily by more than 150 doves, they are reported from nowhere else in our region. The exact location of Trail Wood must be imprinted on a map in their minds. Swift-winged and wide-ranging, they easily may come from as far as forty miles away. Our winter flock may be drawn from an area as large as 160 square miles. Up until a few years ago, even though they maintained their populations with difficulty and in most sections of our state mourning doves are not numerous, these birds were legally shot as game birds. As one of those who testified before the legislative hearings that resulted in these beautiful and often hard-pressed birds' being placed on the protected list, I am delighted to feed the flocks that come to us from so far around each winter.

Four times in every year we see a season ebb and the rising tide of a new season follow. Long before spring replaces winter, we discern in our birds signs of the approaching change. The bills of the starlings lose their blackish hue and turn richly yellow. Those of the evening grosbeaks begin to alter from ivory-yellow to a beautiful shade of almost apple green. The yodeling of the bluejays, the drumming of the downy woodpeckers on dead limbs, the "Phoebee" calling of the chickadees increase as spring draws nearer.

But it is the return of the redwings in early March, each year, that opens the floodgates of change. One day the treetops above our brook are bare; the next they are filled, as though with dense black foliage, with hundreds of homecoming redwings. We open the door to hear what we have looked forward to hearing during all the gales of winter—the commingled voices of the male blackbirds, that tumult of calling, that Niagara of sound, that speaks of all the excitement of spring. In nature's calendar, for us, spring begins when the redwings return. Come wind! Come snow! Spring is within our reach!

In the days that follow, step by step, the sequence of events continues. First the redwings. Then the soft calling of bluebirds over the melting fields. Then, among the wild plums, the white-throated sparrows lifting their voices, so sweet, so pure, in their "Oh, Canada, Canada" song. For me all the frail, enduring beauty of the world finds its voice in the song of the whitethroat. Then comes the wild exulting far-carrying cry of the first flicker. Before breakfast, one morning, we hear the voice of a song sparrow and, one later evening at sunset, the first robin of the year singing in the hickories.

And all the while, our winter birds have been slipping away, drifting north or spreading out into the fields of spring. Suddenly we realize our tree sparrows are gone. So the birds of the winter wind, whose fortunes we follow so long, are replaced by birds of the warm spring breezes. This is the changing of the guard. It is time to prop open the door of the middle shed so the phoebes, coming home, can nest inside.

CHAPTER TWENTY-THREE

The Pasture Rose

Bluets in the meadows, hepaticas in the woods, marsh marigolds in the swamps. So the long parade of the wildflowers gets under way in April. Meadow rue and wild flag, wood anemone and cardinal flower, evening primrose and wood betony, heath aster and pasture rose, they all, and a hundred more, arrive in their appointed sequence. All through the spring and summer and early fall, the gay colors of the wildflowers decorate the green pages of our fields. Just as a century ago, or centuries before that, our flowers follow the same immemorial progression in their blooming.

We have come to know well many of the participants in this annual parade: the wood anemone that closes its flowers at night and the rue anemone that keeps them open; the yellow star flowers of the whorled loosestrife hugging the tops of the leaves; the curious three-lobed dark-red blooms of the wild ginger; the germander with its budding flowerheads as soft as a kitten's tail; the wild indigo with its yellow pealike flowers and its cloverlike leaves; the woodland hepatica, sometimes purple, with—as Richard Cabot once wrote—"the white ends of its stamens shining against its deep purple cup like stars in a summer night." We know the pale-green flowers of the false hellebore, that plant of the lowlands, precocious in growth, premature in death. The rains of April resound, as on a drumhead, on its large green leaves. But, like our Dutchman's breeches—which begins to wither almost as soon as its early blooming is over—the hellebore, unless in flower, turns brown with the rust of age before the spring has hardly ended.

Around the house we have the common cultivated farmyard flowers: iris and phlox and goldenglow, poppies, moss roses, columbine, and a great clump of lilacs half a century old. We have one bush of the Furnessville rose, an old-fashioned sweet-scented pink rose that grew, in my boyhood, at Lone Oak Farm, near Furnessville, on the edge of the Indiana dunes. We have a clump of deep-red peonies from the same source. It is nearly a century old. Beside one wall, a plant not usually thought of for its flowers lifts, each spring, high-piled creamy masses of blooms that suggest cumulus clouds in miniature. This is the rhubarb. Such are the gardens of our farmyard. But beyond the yard, extending away on all sides, are the natural gardens of Trail Wood, the flowers of the fields and woods, wild gardens, comprising all those richly varied blooming plants commonly called weeds.

Yearly, among this living host, we add to our list of wildflower acquaintances. Yearly we discover new and interesting botanical inhabitants of our farm. One summer day, when we were sitting on a rustic bench closed in by ferns and spicebush and overlooking a little cataract in Hampton Brook below my Insect Garden, Nellie noticed an unfamiliar plant nearby. It was about a foot and a half high, with alternating leaves, light-green and lance-shaped, ascending the upright stem to the top where diverging branches held minute yellow-green flowers only about one-sixth of an inch across. But they were formed with the symmetry of jewels. To this plant Carl Linnaeus had given the specific name of *sedoides*, meaning "like sedum." Although it looks more like a goldenrod than a succulent, and is now classified as a member of the saxifrage family, earlier botanists placed it among the stonecrops. The common name by which it is still known is the ditch stonecrop. It is the only representative of its sub-family, *Penthorum*, found in North America. Later in the season, other minute jewels decorate the top of the ditch stonecrop. Then its small seedpods appear, in form as beautiful as five-pointed snowflakes.

In this same damp and shaded spot beside our brook we sometimes come upon a square-stemmed member of the mint family, its hood-shaped blue flowers dangling in rows. This is a plant with an arresting name: the mad-dog skullcap. Venus's looking glass, enchanter's nightshade, mad-dog skullcap—how the names of these inhabitants of our land carry us back to some more primitive, far earlier period in human history! It is on higher, drier, less fertile ground that, in one single place beneath a wild apple tree, we find the wandlike stem of Venus's looking glass, a member of the bellwort family, *Specularia perfoliata*. All up its stem, in blooming time, its blue or violet five-petaled flowers rise from the clasping leaves as though from small green bowls.

In the dusk of summer evenings when we return from the screened-in summerhouse beside the pond, we follow Veery Lane through a cool green cavern beneath a leaning tree. In the dim light of that deeply shaded spot, we encounter the third of these wild plants with arresting names: the enchanter's nightshade, *Circaea lutetiana*. Its generic name is derived from Circe, the enchantress of Greek mythology. Above its leaves, each hairy along the margin, it bears tiny flowers consisting of only two white petals, each deeply notched. By late August the flowers are gone and we return home with our clothes covered with fruit, small rounded burs that anchor themselves in place with a multitude of stiff hairs, hooked at the end. Unlike the deadly nightshade, the enchanter's nightshade is no nightshade at all. Rather it is a member of the evening primrose family.

From time to time, we discover lone flowers, or two or three of a kind, belonging to species that are found nowhere else on our acres. Each such isolated station is a monument to the adventures of a traveling seed. What combination of circumstances, we wonder, established these plants at the spots where we find them?

Near the beginning of the Old Woods Road, we have our only poke milkweeds with their tall stalks and clusters of creamy-white flowers tinged with green. Our single stand of twayblade orchids bloom at one point beside the Old Colonial Road. An odd feature of this orchid is the fact that, unlike other plants with bulbs, its bulb is not buried but is almost wholly above ground. Another lonely plant shines out in the lowland meadow where the woodcock begins its song flights in the spring. For a number of years it has remained there, our only clump of orange milkweed. Remote from all others of its kind, deep in the North Woods, we once saw a single stalk of mullein. How had its seed traveled so far without benefit of parachute or bur?

Varying widely in form are these lonely plants of ours. They range from the goatsbeard, with glints of golden hue playing over the large globe of its dandelionlike seed-head, through the slender three-seeded mercury, that green-flowered member of the spurge family; the curious leafless naked broom rape, pale of stalk and white of flower; and the coral-root orchid, or dragon's-claws, with the waxy white lips of its flowers spotted with crimson. Some of these isolated plants come and go. We may see them once or twice and then no more. Others, like the goatsbeard and the orange milkweed, remain for years.

Sprung from some wind-borne seed and growing alone one summer, a tall blue lettuce beside the road at the entrance of our lane continued to shoot upward until it set an all-time record for the height attained by one of our herbaceous plants. The blue-tinted flowers that appeared at the top of its stalk bloomed eleven and a half feet above the ground. Other wildflowers of ours have attained uncommon heights. One mullein stalk that ascended among the branches of a pear tree reached a height of six feet, eleven and a half inches. The goldenrod *altissima* often rises above five feet, higher than Nellie's head. And, not far from the bench where we saw the ditch stonecrop, a jack-in-the-pulpit, one year, kept lifting above the moist rich soil until it stood forty-one inches, almost three and a half feet, above the ground.

In contrast to the wildflowers that attract attention because of their rarity or size, there are those that stand out in the mass, that comprise great plant communities, that thrive and multiply and spread. In our pastures these appear in a season-long sequence of dominant blooms—the pale sheets of the little bluets, the enameled yellow of the buttercups, the light rose-purple of the Philadelphia fleabane—each of its flowers formed of a hundred or more slender rays. So, as the floral wheel of summer revolves, the daisies, the asters, the goldenrod spread away in waves of color.

During April, that time when spring itself expands like a flower, that time of a hundred firsts—the first marsh marigold, the first rue anemone, the first trillium, the first bloodroot, the first coltsfoot—Nellie and I always find a special pleasure in the multitudinous return of the smallest of the pasture flowers, the little bluets. Known by many names—innocence, Quaker ladies, nuns, blue-eyed babies—these demure four-petaled blooms, pale blue or pale lavender, occasionally white, mass together in patches yards across. They are at their best when cows keep the meadows cropped short. From a distance, they stand out, pools of delicate color amid the rich green of the early grass.

In the midst of hundreds of these flowers, one day, my eye by chance caught sight of one different from the rest. Among all those normal four-petaled blooms this one flower had five petals. From time to time we encounter such oddities in wandering among our wildflowers. Once it was a yarrow with pink instead of white florets. I suspect some unusual element in the soil where it grew accounted for its abnormal color. Beside Stepping Stone Brook, another time, in the midst of a large clump of brilliant cardinal flowers we noted one stalk of pure white blooms—an albino cardinal flower. A friend of ours, a mile or two from Trail Wood, once picked a black-eyed Susan with petals that were green instead of yellow. And in a small hollow in our South Woods we come upon clumps of Indian pipes that, instead of being waxy ghostly white, are a beautiful shade of pink, sometimes with the color even deepening into red.

Bee pastures and butterfly pastures, nectarfields for insects, are provided by many of our plant communities such as the stretches of frosty-green narrow-leaved mountain mint and the golden islands of the rough-fruited cinquefoil. There, all through the long warm days, the air is filled with humming. For bees as well as butterflies, the nectar of the mountain mint appears to be a favorite food. I find honeybees, bumblebees, carpenter bees, digger wasps, paper-making wasps, and butterflies of many kinds—coppers, hairstreaks, fritillaries, dark little skippers, blue-eyed graylings—flocking to the small white flowers. Sometimes I come upon a dozen pearl crescent butterflies all feeding at a single clump. And here too, hidden among the flowers, is an insect that has no interest in nectar. About three-eighths of an inch long and clad in camouflaging greens and yellows, the ambush bug lies in wait with its powerful grasping forelegs and its sharp sucking beak. The victim that comes within reach is quickly seized and subdued and its vital fluids are drained away to nourish the little predator. In a single clump of mountain mint I once counted fifty-four ambush bugs. On another occasion, in an island of these plants less than a dozen paces across, a census of such predators revealed a total of 158.

At the opposite extreme from the narrow-leaved mountain mint, swarming with insects, stands the brilliant cardinal flower. It blooms, sometimes massed more than a hundred together, along Hampton Brook. There it is virtually ignored by the nectar-gathering insects. So far as we can see, it is fertilized almost entirely by ruby-throated hummingbirds. Each summer we watch these birds as they hover in a shimmer of metallic colors beside the red lobelias, skillfully running their slender bills between the petals to reach the unseen store of nectar within.

Before I leave the subject of our plant communities, three others should be mentioned—one near the Brook Crossing in the North Woods, the second along the south side of a stone wall that runs east and west, the third near the edge of a woodland on the trail to Bracken Hill. In the North Woods, where the oaks and maples gradually descend to the edge of a swamp, the slope provides our greatest concentration of pink lady's-slippers. There we sometimes stand with more than thirty of these largest of our orchids around us. Along the south side of the east-and-west wall, where the spring sun brings early warmth, our bloodroot colony flourishes. Sixty or more bloom at once, the deeply cut pale-green leaves embracing the stems like hands holding aloft the shining white stars of the flowers. And along the path to Bracken Hill, where the little Canada mayflower grows, we encounter our principal garden of wild geraniums. With five parts to the leaves and five rose-purple petals to the flowers, between 100 and 200 bloom together in the latter weeks of May. I rarely pass this woodland garden without recalling how Winston Churchill, during the Second World War, observed that a love of flowers is one of the noblest qualities of the human mind. He added:

"You can live a long time with a geranium."

One summer, that grand old man of wildflower photography, Samuel H. Gottscho, visited us with his camera. He was then in his ninetieth year. After adding to his immense collection of flower portraits that day, he made a remark that has remained in my mind. It seemed a high compliment for the fields of Trail Wood. Gazing about him at our wild unplanted gardens as he was leaving, he observed:

"This is the kind of place where a humble weed can raise its head and be admired."

Aside from man's aesthetic response and the utilitarian interest of the nectar and pollen gatherers, few creatures pay much attention to wildflowers. In the light of this a strange thing happened, some years ago, on a farm here in northeastern Connecticut. Among a herd of young steers in one of Lou Chatey's pastures in Ashford, one, light gray, almost silvery in color, was outstandingly beautiful. It alone, among all the animals in the field, seemed attracted by wildflowers. Every day it would spend part of its time carrying flowers about in its mouth without eating them. Often it would be engaged in this manner for half an hour at a time. On some occasions, it would lay down the flowers, eat grass for a while, then pick them up again and carry them about as before. How can we explain its singular behavior? What was the source of its special interest in field flowers? This is the only instance of the kind that I have encountered.

Have you ever wondered what a hypothetical Man from Mars, arriving on earth, would find most interesting? Would it be our skyscrapers? Our color TV? Our superjets? Perhaps not. I remember the story once told me by a nurse who had spent a year at a hospital at

Point Barrow, at the northernmost tip of Alaska. At Christmastime a group of Eskimo children were invited to a party on the second floor. For the first time in their lives, they saw a decorated and lighted Christmas tree. For the first time they received gifts from a Santa Claus. For the first time they heard Yuletide carols and ate a Christmas dinner. But none of these things had any great interest for one small boy. The only thing he wanted to do, all day long, was stand by the windows looking out and down. Never before had he been in a building with a second story. Viewing the world from this new and elevated position was the one thing that absorbed his attention. So, I suspect, the things that would most interest our imaginary Man from Mars might be none of our complicated advances in science, but rather things usually overlooked, simple things, those commonplace everyday miracles such as the growing of the green grass, the rippling of the little brook, the blooming of the pasture flower.

Year by year we watch the advance and retreat of our different wildflowers. How our six kinds of violets, our nine kinds of orchids, our five kinds of hawkweeds, our six kinds of St. Johnswort, the blue flag and the creamy-white bellwort, the mullein and the meadow rue —called "bobolink weed" in an earlier New England day—fare in any given season depends on many factors. Hard times for one species are good times for another. In years of drought we see some plants forge ahead; in times of abnormal rainfall others advance. Even when the annual rainfall is normal that is not the whole story. *When* it rains, as well as *how much* it rains, is important to the individual plant. If it rains today instead of tomorrow, the history of something is changed. A shower—at a certain time—a series of chill days—at a certain time—and the life story of a plant is altered. A complex combination of circumstances, in a constantly shifting balance, governs the fate of the individual plant. Yet the wildflower is a supreme example of beauty combined with strength. There is strength in the frailest flower, strength in the continuity of its kind. Refined by evolution, nature's products are enduring and tough. The weed outlasts the hoer and the hoe.

In our annual succession of wildflowers at Trail Wood, early September brings the high noon of the asters and the high tide of the goldenrod. In those far-off days of the seventeenth century when William Wood published his *New England's Prospect*, he could write of "the wasp" and "the grasshopper." So, in those days, it was probably "the aster" and "the goldenrod." But in the years that have intervened botanists have been busy classifying species, dividing and splitting, until Gray's *Manual of Botany* now lists more than 130 kinds of goldenrod for the Central and Northeastern states. Almost one-tenth of these we have found growing on our land. And of the nearly 150 kinds of asters Gray lists, nearly one-eighth of them are native to Trail Wood.

Eighteen is the total number of our asters. The list includes the large-leaved, the wavy-leaved, the smooth, the stiff, the willow, the New England, the heart-leaved, the white wood, the purple-stemmed, the toothed white-topped, the New York, the small white, the common blue wood, the sharp-leaved, the white or mountain aster, the starved, the white heath, and the late purple. If we count the silverrod, we have an even dozen goldenrod. The others are the tall goldenrod, *altissima*; the early goldenrod, *juncea*; the late goldenrod, *gigantea*; the rough-stemmed or wrinkle-leaved goldenrod, *rugosa*; the Canada goldenrod, *canadensis*; the gray goldenrod, *memoralis*; the elm-leaved goldenrod, *ulmifolia*; the late-blooming blue-stemmed or wreath goldenrod, *caesia*; the showy or noble goldenrod, *speciosa*; the lance-leaved goldenrod, *graminifolia*; sometimes called the fragrant; and the bog or swamp goldenrod, *uliginosa*, which grows as far north as New Brunswick. The rough-stemmed *rugosa* is the dominant goldenrod of our fields.

Across acres of old pastures, where the goldenrod are taking over, late August and early September turn the land into a tumbling sea of the richest yellow. The fields are awash with waves of goldenrod that flow across the slopes and break against the stone walls and the woods. Dusty, golden, polleny goldenrod—what a host of insect life revolves around its yellow masses! What bee pastures it provides! Over these September fields monarch butterflies drift southward, beginning their long migration of fall. As many as eight or ten at a time sometimes flutter down to cling to a single clump of the noble or showy goldenrod. For minutes they hang there motionless, drinking the rich nectar at this floral way station along their route.

Another favorite of the butterflies, in these waning days of summer, is the white heath aster. For us, an event of the season, each year, is the coming of these shining mounds of frosty white to the slope that descends from the house to the pond. So dense are the masses of its blooms that the plants often seem mounded over with soft new-fallen snow. One of the common names of the heath aster is farewell-summer. Autumn is in the air, the year of the flowers is nearing an end, when this late-blooming aster comes to the hillside.

All through the season of the wildflowers Nellie and I go afield accompanied by a pocket magnifying glass. Through its lens we see tiny designs, minute details, unsuspected features of even the most diminutive of the blooms expanding before our eyes. Have you ever enjoyed examining thus the delicate sculptured striations on the overlapping parts of a hop clover bud or the central pattern, rich as gold, at the heart of the flower of the small blue-eyed grass or the ten-petaled jewels—each petal ribbed and waxy-white—that star the filmy clouds of branching stems where the stitchwort sprawls among the higher pasture grass?

In this adventure in a Lilliputian realm, you soon note that there is little relationship between the size of the flower and the intensity of its perfume. Some of the smallest blooms—such as those of the rough bedstraw—fill the air for a long way around them with the sweetness of their scent. I have to hold the spotted pipsissewa close to my nose to catch the faint, pleasant perfume of its fleshy flowers. But when I come within a dozen yards of the sweet clethra I am enveloped in its heavy fragrance.

Neither faint nor overpowering, but perfectly suited to its character, is the fragrance of another flower that I always encounter with delight in the fields, that old friend of mine, the pasture rose, *Rosa carolina*. An inhabitant of dry or rocky soil, of old abandoned hillsides, it ranges from Newfoundland to Louisiana. It is the commonest of all our wild roses. For me, among all the colors spread across the fields at the end of spring, its clear, unmuddied, modest hue is one of the most appealing. These pink flowers, two or three inches across, decorate bushes that are sometimes no more than a foot or two high. The common names of the pasture rose include the low or dwarf wild rose. Its flowers are frequently few or solitary. But those blooms, with their four clear-pink overlapping petals, their delicate perfume, their setting amid dusty fields or rocky slopes, possess an unassuming beauty that produces a lasting impression. Examine one of the petals beneath a magnifying glass and you discover unsuspected beauty, fine lines or veins of darker pink that radiate upward from the base.

Thinking of all that host of varied forms and hues, the wildflowers of the spring and the summer and the early fall, I sometimes try to pick a favorite. I never succeed. I have no single favorite. But there are scores that qualify as "one of my favorites." However, it is curious how often in my mind's eye I see one simple form, one clear-pink hue, one golden circle of anthers, among nature's rainbow of wildflowers. What I recall is the quiet, serene beauty of the little pasture rose. Like the unpretentious song of the field sparrow in the hush of sunset, the pasture rose epitomizes, for me, the peace of evening at the end of the long summer's day. It symbolizes the everyday satisfactions of this country life of ours. At such moments we become keenly aware of how much of the sweetness of that life depends on simple things, such simple things as the song of the field sparrow, the flower of the pasture rose.

CHAPTER TWENTY-FOUR

Faraway and Near-at-Hand

When I stepped out under the rising sun of this June daybreak, the air was honey-scented, the grass all glittering with dew, the world glowing and green and alive. Fifteen times, while we have been here, June has returned to Trail Wood. Although small natural changes in our surroundings have been innumerable, the general scene remains the same. The fields and woods and brooks are as we first saw them.

During this decade and a half, while we have become increasingly familiar with our trails, our swamps, the fern banks, the brooks, and the far corners of our land, we have experienced surprisingly little feeling of possession, of ownership. The deed is ours. But in our minds the trees own the land almost as much as we do. We are one with the robin among the wild cherries in the sunshine. We own Trail Wood together, we and every other wild creature that shares with us these country days.

In this pleasant land we are—like the cottontail rabbit feeding on clover, the goldfinch among the thistle seeds, the green frog half-submerged at the pond's edge in the August heat—a transitory dweller making the most of each passing day. In our minds the sense of guardianship, of responsibility, surpasses that of ownership. Our title to the farm gives us the power to protect this small fragment of the earth, these woods, these streams, these fields, and all the inhabitants they contain.

On this June morning, high overhead, a jetliner toils from east to west trailing a thin straight line of black behind it. That slender thread of descending soot is the only visible sign of pollution in our air. But the things we have enjoyed so intensely here—the quiet, the pure air, the space, the privacy, unchanged nature around us—all of these, even while we have been enjoying them, have become rarer, have become more nearly luxuries in the world.

All through earlier ages, primitive men, few in number, hemmed in by dark forests, facing the vastness of the seas, stood on the defensive in the presence of nature. It was Nature against Man. Now, in a great reversal that has come almost suddenly to a culmination, it is Man against Nature. No previous dweller on the planet possessed such power to alter and disrupt and destroy. Looking about me on such a morning as this, it is difficult to realize that beyond the horizon, even beyond the seas, on the other side of the world, forces are at work that may blight or lay waste such tranquil scenes as these. The explosion of an atom bomb thousands of miles away increased the strontium-90 content of grass in New England pastures and in the milk of New England cows. The faraway threatens the near-at-hand. Our concern for nature and our concern for our fellow men are bound together. The welfare of city and country alike ultimately depends upon the solution of the same great problems.

During these years while we have lived at Trail Wood, another change, a counterbalancing movement, has gained momentum. Destruction continues. But no longer does it continue unchallenged. Wildlife stands less alone. The welfare of all nature has assumed greater importance. To the lonely voices of the Henry Thoreaus and John Muirs of the past, voices crying in the wilderness *for* the wilderness, singing the praises of wildness for its own sake, and speaking up for the rights of lesser creatures, has been added a host of new voices, the voices of a generation more concerned, more aware.

I suppose in every age men feel some evil spirit is abroad. They see the lid of Pandora's box always rising. Richard Jefferies, the English nature writer, in another century wrote: "There is a sense of uncertainty in the atmosphere of the age: no one can be sure that the acorns he plants will be permitted to reach their prime." With us, at Trail Wood, the greater the threat of the faraway to the near-at-hand, the more precious grows this green and pleasant foothold on the earth. Each morning we breathe what the urban man thinks of as "that wonderful vacation air." Here we have found the simple, the good, the satisfying life—not for everyone, perhaps, but certainly for us. No other time, no other place would suit us better.

Different people with different interests would have recorded different things about our life on this old farm. Given our outlook and our interests, it has been our closeness to nature, our daily existence on the edge of wildness that has made the most profound impression. Here we bought sunrises and violets and whippoorwills as well as woods and pastures. If you wonder if this life's original sweetness did not wear away as time went on, if this life did not become more tame and dull with closer acquaintance, I have news, and the news is good. After a decade and a half, this life is still as satisfying, still as near the heart's desire, the last minute as fine as the first. Our acres remain filled with freshness and surprise as though we were visitors, newcomers, rather than long-time owners of the land.

We did not come to Trail Wood, in the first place, to analyze country living, to examine and dissect it, but rather to enjoy and appreciate it. This book you have been reading is not intended as a guide to rural living; it is an account of a naturalist—two naturalists—who bought an old farm and something of the adventures and satisfactions found there. In the main it has been a book of pleasant times remembered, times we would gladly live again.

Sitting under the apple trees, walking down the lane, following the wood trails, circling the pond at sunset, our life here has seemed all kernel and no husk. It embraces one of the rarest things in modern life—moments of solitude. Some people need them more than others, Nellie and I more than most. For me, I never feel in some great urban center, some New York or Chicago or Los Angeles, as I do at Trail Wood: "*This* is my own, my native land!" Time, which has no beginning and no end, has an end for all our beginnings. But for those who find their pleasure out-of-doors, who have known the enchantment of "the dust, the sunshine and the rain," all the years of existence represent a long love affair with the earth, *this* earth, the only earth we know.

INDEX